Fundamentals of Astrophysics

This concise textbook, designed specifically for a one-semester course in astrophysics, introduces astrophysical concepts to undergraduate science and engineering students with a background in college-level, calculus-based physics. The text is organized into five parts covering: stellar properties; stellar structure and evolution; the interstellar medium and star/planet formation; the Milky Way and other galaxies; and cosmology. Structured around short easily digestible chapters, instructors have flexibility to adjust their course's emphasis as it suits them. Exposition drawn from the author's decade of teaching his course guides students toward a basic but quantitative understanding, with "quick questions" to spur practice in basic computations, together with more challenging multipart exercises at the end of each chapter. Advanced concepts such as the quantum nature of energy and radiation are developed as needed. The text's approach and level bridges the wide gap between introductory astronomy texts for nonscience majors and advanced undergraduate texts for astrophysics majors.

Stan Owocki is a professor at the Department of Physics and Astronomy at the University of Delaware, following positions at Harvard and University of California San Diego. He has coauthored more than 300 scientific papers, with his research focusing on mass loss from luminous, massive stars. His teaching at undergraduate to graduate levels includes the development of his flagship "Fundamentals of Astrophysics" course, which forms the basis for this textbook.

Fundamentals of Astrophysics

STAN OWOCKI

University of Delaware

CAMBRIDGE
UNIVERSITY PRESS

University Printing House, Cambridge CB2 8BS, United Kingdom

One Liberty Plaza, 20th Floor, New York, NY 10006, USA

477 Williamstown Road, Port Melbourne, VIC 3207, Australia

314–321, 3rd Floor, Plot 3, Splendor Forum, Jasola District Centre, New Delhi – 110025, India

79 Anson Road, #06–04/06, Singapore 079906

Cambridge University Press is part of the University of Cambridge.

It furthers the University's mission by disseminating knowledge in the pursuit of education, learning, and research at the highest international levels of excellence.

www.cambridge.org
Information on this title: www.cambridge.org/9781108844390
DOI: 10.1017/9781108951012

First published 2021

A catalogue record for this publication is available from the British Library.

ISBN 978-1-108-84439-0 Hardback
ISBN 978-1-108-94812-8 Paperback

Contents

Preface

This book grew directly out of class notes developed from over a decade of teaching a one-semester course of the same title to second- and third-year undergraduate science and engineering majors at the University of Delaware (UD). Although structured within UD's Department of Physics and Astronomy, in which I am a faculty member, generally only about a third of the *c.* 30 students in the class are physics or astronomy majors; others major in engineering (mechanical and chemical), biology (including biophysics), chemistry, and computer science. The main prerequisite is a year of college-level, calculus-based physics, along with associated math courses. There is *no* presumption of prior, direct study of higher-level physics such as relativity, quantum physics, electricity and magnetism, and thermodynamics. Indeed, there is no presumption of prior study in astronomy, not even from a descriptive "Astronomy for Poets" course. This and such upper-level physics concepts are introduced as needed.

Grounded in development for that course, this text is specifically targeted to be a bridge between the copious number of introductory astronomy texts aimed mainly at nonscience majors, and the handful of astrophysics books aimed at upper-level physics majors, which thus assume a background well beyond just first-year physics. Its moderate length, moreover, also distinguishes this from books that are at a similar level, but are so extensive as to be more suited as a reference than an instructional text.

Within this context, the goal of this book is to help guide such students to a broad understanding of the *Fundamentals of Astrophysics*, ranging from basic properties of stars to the principles of Big-Bang cosmology, all within the structure of a one-semester course. The aim for coverage within a single semester naturally presents some significant challenges, for both the students and the instructor. Over more than a decade as an instructor, I have in various semesters experimented with de-emphasizing some topics, sections, and even chapters, in favor of others. For instance, as the course notes developed to include areas of increasing topical interest, e.g., the discovery and modeling of the ever-growing number of exoplanets, it has been necessary to de-emphasize or even skip other chapters or topics, for example streamlining coverage on galaxies to focus on the Hubble expansion with the goal of getting through at least to the scale-factor and Cosmic Microwave Background (CMB) sections of the cosmology chapters.

This need for flexibility in instructor choice was the main factor setting the book's basic structure, with a large number (33) of relatively short (∼5–10 pages) chapters that themselves are broken up into ∼1–3 page "sections." In general, I find most chapters can be covered in one or two lectures, and some shorter ones can even be combined into a single lecture.

The chapters are, in turn, organized into five "parts." The longest of these is Part I on the basic properties of stars, which aims to provide students with an understanding of how we are able to determine their physical properties from measurements of mere points of light; it includes 14 relatively brief chapters, concluding with one on the Sun. (Although this provides a reality check on our idealized portrait of stars as static spheres of gas, I often skip it for brevity, as well as other less-central topics such as stellar rotation.) Part II then reviews stellar structure and evolution within the framework of single stars. Less basic aspects such as evolution of interacting binaries are deferred, e.g., to a student exercise on the Algol paradox.

Part III then links discussion of the interstellar medium (ISM) with formation of stars and planetary systems. The last is a burgeoning topic of ever-growing scientific and even public interest, and the aim here is to give a basic overview of the detection and modeling methods. Part IV next extends the discussion to our Milky Way and other galaxies, including the Hubble expansion of the universe. This leads naturally to the Part V review of Big-Bang cosmology, grounded within a Newtonian treatment of the expansion, and the associated formation of the CMB. The main text concludes with a chapter on the eras of the early universe, including the notion of an early era of rapid inflation.

To supplement this five-part narrative grounded in astrophysics, four appendices summarize key background physics topics on: (1) atomic structure; (2) excitation and ionization; (3) opacity; and (4) radiative transfer.

In keeping with the broad educational mantra that students learn best by doing, each chapter ends with a section containing questions and exercises. The former are relatively short and focus on one or two concepts that the students should be able to answer relatively quickly once they have read and studied the chapter; I often use selected questions for in-class discussions or quizzes, to encourage and test students on doing the assigned reading (which frankly can be quite a challenge).

The exercises are longer, with multiple parts, and are generally intended for weekly homework assignments. Some are still pretty straightforward "plug-ins" to chapter formulae; but others are intended to be more challenging and thought-provoking, in some cases even introducing extensions not directly covered in the text (e.g., the "Ledoux criterion" in Exercise 4 of Chapter 17). Others aim to connect astrophysical concepts to student major interests, e.g., the "space elevator" (Exercise 6 in Chapter 10), which relates gravitational binding of orbits to mechanical engineering. The very last exercise (Exercise 5 in Chapter 33) directs the students to research the concept of a "multiverse," and discuss whether this even constitutes a scientific theory.

Finally, while this book was developed and written for a one-semester course, it could readily also serve as the core text for a two-semester sequence, allowing for a more leisurely and in-depth coverage of the full breadth, supplemented perhaps by linkage to related advanced topics, such as those listed in the instructor resources, which can be found at www.cambridge.org/owocki.

Part I

Stellar Properties

1 Introduction

1.1 Observational versus Physical Properties of Stars

What are the key physical properties we can aspire to know about a star? When we look up at the night sky, stars are just little "points of light"; but if we look carefully, we can tell that some appear brighter than others, and moreover that some have distinctly different hues or colors. Of course, in modern times we now know that stars are really "suns," with properties that are similar – within some spread – to those of our own Sun. They appear much much dimmer only because they are much much farther away. Indeed they appear as mere "points" because they are so far away that ordinary telescopes can almost never actually resolve a distinct visible surface, the way we can resolve, even with our naked eye, that the Sun has a finite angular size.

Because we can resolve the Sun's surface and see that it is nearly round, it is perhaps not too hard to imagine that it is a real, physical object, albeit a very special one, something we could, in principle, "reach out and touch." (Indeed a small amount of solar matter can even travel to the vicinity of the Earth through the solar wind, coronal mass ejections, and energetic particles.) As such, we can more readily imagine trying to assign values of common physical properties – e.g., distance, size, temperature, mass, age, energy emission rate, etc. – that we regularly use to characterize objects here on Earth. Of course, when we actually do so, the values we obtain dwarf anything we have direct experience with, thus stretching our imagination, and challenging the physical intuition and insights we instinctively draw upon to function in our own everyday world. But once we learn to grapple with these huge magnitudes for the Sun, we then have at our disposal that example to provide context and a relative scale to characterize other stars. And eventually as we move on to still larger scales involving stellar clusters or even whole galaxies, which might contain thousands, millions, or indeed billions of individual stars, we can try at each step to develop a relative characterization of the scales involved in these same physical quantities of size, mass, distance, etc.

So let us consider here the properties of stars, identifying first what we can directly *observe* about a given star. Since, as already noted, most stars are effectively a "point" source without any (easily) detectable angular extent, we might summarize what can be directly observed as three simple properties.

1. **Position on the sky:** Once corrected for the apparent movement due to the Earth's own motion from rotation and orbiting the Sun, this can be characterized by two coordinates – analogous to latitude and longitude – on a "celestial sphere." Before modern times, measurements of absolute position on the sky had accuracies on order of an arc minute (1/60 of a degree, abbreviated here as arcmin); nowadays, it is possible to get down to a few hundreths of an arc second (1/3600 of a degree, abbreviated here as arcsec) from ground-based telescopes, and even to about a milliarcsecond ($\sim 10^{-3}$ arcsec, abbreviated here as mas) (or less in the future) from telescopes in space, where the lack of a distorting atmosphere makes images much sharper. As discussed in Section 2.2, the ability to measure an annual variation in the apparent position of a star due to the Earth's motion around the Sun – a phenomena known as "trignonometric parallax" – provides a key way to infer distances to at least the nearby stars.

2. **Apparent brightness:** The ancient Greeks introduced a system by which the apparent brightness of stars is categorized in six bins called "magnitude," ranging from $m = 1$ for the brightest to $m = 6$ for the dimmest visible to the naked eye. Nowadays we have instruments that can measure a star's brightness quantitatively in terms of the energy per unit area per unit time, a quantity known as the "energy flux" F, with units erg/cm^2/s in CGS or W/m^2 in MKS. Because the eye is adapted to distinguish a large dynamic range of brightness, it turns out its response is *logarithmic*. And since the Greeks decided to give dimmer stars a higher magnitude, we find that magnitude scales with the log of the *inverse* flux, $m \sim \log(1/F) \sim -\log(F)$, with the $\Delta m = 5$ steps between the brightest ($m = 1$) to dimmest ($m = 6$) naked-eye stars representing a *factor 100 decrease* in physical flux F. Using long exposures on large telescopes with mirrors several *meters* in diameter, we can nowadays detect individual stars with magnitudes $m > +21$, representing fluxes a million times dimmer than the limiting magnitude $m \approx +6$ visible to the naked eye.

3. **Color or "spectrum":** Our perception of light in three primary colors comes from the different sensitivity of receptors in our eyes to light in distinct wavelength ranges within the visible spectrum, corresponding to red, green, and blue (RGB). Similarly, in astronomy, the light from a star is often passed through different sets of filters designed to transmit only light within some characteristic band of wavelengths, for example the UBV (ultraviolet, blue, visual) filters that make up the so-called Johnson photometric system. But much more information can be gained by using a prism or (more commonly) a diffraction grating to split the light into its spectrum, defining the variation in wavelength λ of the flux, F_λ, by measuring its value within narrow wavelength bins of width $\Delta\lambda \ll \lambda$. The "spectral resolution" $\lambda/\Delta\lambda$ available depends on the instrument (spectrometer) as well as the apparent brightness of the light source, but for bright stars with modern spectrometers, the resolution can be 10 000 or more, or indeed, for the Sun, many millions. As discussed in Section 6.1, a key reason for seeking such high spectral resolution is to detect "spectral lines" that arise from the absorption and emission of radiation via transitions between discrete energy levels of the atoms within the

star. Such spectral lines can provide an enormous wealth of information about the composition and physical conditions in the source star.

Indeed, a key theme here is that these three apparently rather limited observational properties of point-stars – position, apparent brightness, and color spectrum – can, when combined with a clear understanding of some basic physical principles, allow us to infer many of the key physical properties of stars, for example:

1. **Distance**
2. **Luminosity**
3. **Temperature**
4. **Size** (i.e., radius)
5. **Elemental Composition** (denoted as X, Y, Z for mass fraction of H, He, and of heavy "metals")
6. **Velocity** (both radial [toward/away] and Transverse ["proper motion" across the sky])
7. **Mass** (and surface **gravity**)
8. **Age**
9. **Rotation** (period P and/or equatorial rotation speed $V_{\rm rot}$)
10. **Mass Loss Properties** (e.g., rate $\dot{M}$ and outflow speed V)
11. **Magnetic Field**

These are ranked roughly in order of difficulty for inferring the physical property from one or more of the three types of observational data. The list also roughly describes the order in which we will examine them in the remaining chapters of Part I. In fact, except for perhaps the last two, which we discuss only briefly (though they happen to be two specialities of my own research), a key goal is to provide a basic understanding of the combination of physical theories, observational data, and computational methods that makes it possible to infer each of the first nine physical properties, at least for some stars.

1.2 Powers-of-Ten Scale Steps from Us to the Universe

Before proceeding, let us make a brief aside to discuss ways to get our heads around the enormous scales we encounter in astrophysics.

As illustrated in Figure 1.1, one approach is to use a geometric progression through *powers of ten*,[1] from the scale from our own bodies, which in standard metric (MKS) units is of order 1 meter (m), to the progressively larger scales in our universe.

For example, the meter itself was originally *defined* (in 1793!) as one ten-millionth, or 10^{-7}, of the distance from Earth's equator to poles; this thus means a total of *seven* steps in powers of ten from the human scale to that of our Earth. This is the largest

---

[1] There are many online versions, including a rather dated (1977) but still informative movie titled *Powers of Ten*, which you can readily find by Google; for a modern version, see www.htwins.net/scale2/.

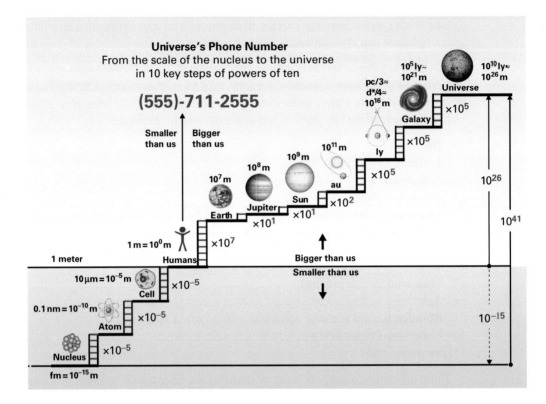

Figure 1.1 Graphic to illustrate key powers-of-ten steps between our own human scale of 1 m, both upward to the scale of the universe (10^{26} m), and also downward to the scale of an atomic nucleus (10^{-15} m). As a mneumonic, this is cast as a 10-digit "telephone number," with the 3-digit "area code" representing the three steps of 10^{-5} from us down to the nucleus, and 7-digit main-number representing seven key steps to the scale of the universe.

scale for which most of us have direct experience, e.g., from overseas plane travel, or a cross-country drive.

The other, rocky inner planets are somewhat smaller but of the same order as Earth; among the outer, gas giant planets Jupiter is the largest, about a factor 10 larger than Earth, while the Sun is about another factor 10 larger still, with a diameter $D_\odot \approx 1.4 \times 10^6$ km, about a factor 100 bigger than Earth, or of order 10^9 m.

The Earth–Sun distance, dubbed an "astronomical unit" (au), is about 100 solar diameters, at 150 million km. This is of order 10^8 km $= 10^{11}$ m, or four further powers of ten beyond the scale of our Earth, and so a total of 11 orders of magnitude bigger in scale than our own bodies.

An alternative way to characterize this is in terms of the time it takes light, which propagates at a speed $c = 300\,000$ km/s, to reach us from the Sun; a simple calculation gives $t = d/c = 1.5e8/3e5 = 500$ s, which is about 8 minutes; so we can say the Sun is 8 *light minutes* from Earth.

By contrast, it takes light from the next nearest star, Proxima Centauri, about *four years* to reach us, meaning it is at a distance of 4 light *years* (ly). A simple calculation shows that one year is 1 yr = $365 \times 24 \times 60 \times 60 \approx 3 \times 10^7$ s; so multiplying by the speed of light $c = 3 \times 10^5$ km/s gives that 1 ly $\approx 9 \times 10^{12}$ km, or of order 10^{16} m. Thus, the scale between the stars is another 5 orders of magnitude greater than that of the Earth–Sun distance, or 16 orders greater than that of ourselves.

The Sun is only one of about 100 billion (10^{11}) stars in our Milky Way Galaxy, a disk that is about 1000 ly thick, and about 100 000 ly across. Thus, our Galaxy is another 5 orders of magnitude bigger than the scale between individual stars, or about 10^{21} m, thus 21 orders bigger than us.

The universe itself is about 14 billion years old (14 Gyr), meaning that the most distant galaxies we can see are of order 10^{10} ly $\approx 10^{26}$ m away. We thus see that 26 powers of ten takes us from our own scale to the scale of the entire observable universe!

To recap, seven powers-of-ten steps take us from human scale to the Earth; then powers-of-ten steps of 1, 1, and 2 take us from the Earth to the size of Jupiter, Sun, and the Earth–Sun distance. Then three successive power-of-ten steps of 5 take us to the distance of the nearest other star; to the size of our Galaxy; and finally to the size of the universe. It can be helpful to remember this 711-2555 rule as a mnemonic – like a 7-digit telephone number – to capture the progression between key scales that characterize our place in the universe.

Indeed, we can extend this even to *small* scales, by noting that five powers of ten *smaller* takes us successively to the characteristic size of a cell, 10^{-5} m = 10 micron (μm); then to the size of atoms, 10^{-10} m = 0.1 nanometer (nm); and finally to the scale of an atomic nucleus, 1 femtometer (also known as "fermi") or 1 fm = 10^{-15} m.

The full sequence of steps over this span thus looks something like a 10-digit phone number with area code: 555-711-2555, representing the powers-of-ten steps from scales of nuclei to atoms to cells to us to Earth to Jupiter to Sun to astronomical unit (distance to Sun) to light year ($\sim$ distance between stars) to our Galaxy to the universe.

In addition, the enormous timescales at play in the universe can likewise be difficult to grasp.

As illustrated in the top panel of Figure 1.2, humans experience time in our everyday world on the scale of a second, which is roughly the order of a single heartbeat. We live a maximum of about 100 years, or about 3 *billion seconds*. In comparison, it is estimated that the Earth is about 4.4 billion *years* old, almost as old as the Sun and the rest of the Solar System. The Sun is expected to sustain its current energy output for about another 5 billion years, and so have a full lifetime of about 10 billion years. And as discussed in Chapter 8, the lifetimes of other stars can depend strongly on their mass; the most massive stars (about 100 solar masses) live only about 10 *million* years, while those with mass less than the Sun are expected to last for up to 100 billion years, much longer than the current age ($\sim$14 Gyr) of the universe!

Finally, the bottom panel of Figure 1.2 gives a similar graphic for the range of speeds, from our own slow walk, through others (bicycles, cars, airplanes) we experience, then ranging to speeds of the Moon, Earth, and Sun in their orbits, to stellar winds and supernovae, and finally ending with the maximum possible speed, the speed

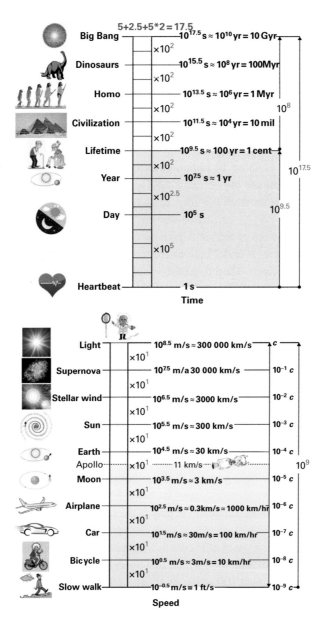

Figure 1.2 Graphics to illustrate the human-to-universe range of scales for time (top) and speed (bottom). The gray highlights the ranges of direct human experience.

of light, $c = 3 \times 10^8$ m/s. The right-hand axis relates the fraction of the light speed for each of the progression of nine powers from walking to light itself.

The following chapters explain how we are able to discover these fundamental properties of stars, beginning with their distance.

1.3   Questions and Exercises

Quick Questions

1. Discuss how we can estimate the temperature of a warm or hot object, e.g., a stove, without touching it.
2. Discuss the ways we estimate distances and sizes in our everyday world.
3. Discuss what sets our perception limits on the smallest intervals of time. How might this differ in creatures of different size, e.g., a fly versus a human?
4. For a typical car highway speed of 100 km/hr, about how long does it take to travel from coast to coast? How does this compare to how long it would take to drive to the Moon? To the Sun? To Alpha Centauri?
5. Why might astronomical observations be useful in measuring the speed of light?
6. How does the speed of sound on Earth compare with the speed of light?
7. About how old is the oldest living thing on Earth? What about the oldest animal?

Exercises

1. *Speeds*
   a. At what speed does a person at the equator move due to the Earth's rotation? Give your answer in mi/hr, km/hr, and m/s.
   b. What is the speed of the Earth in its orbit around the Sun? Give your answer in au/yr, km/s, mi/hr, and in terms of the fraction of the speed of light v_{orb}/c?
   c. The Sun is about 28 000 ly from the center of the Milky Way, and takes about 200 million years to complete one "Galactic year." What is the speed of the Sun in its orbit around the Milky Way, in km/s. In ly/yr? In terms of the fraction of the speed of light v_{orb}/c?
2. *Sun*
   The Sun has a radius of about 700 000 km.
   a. How many solar radii in 1 au? In 1 ly?
   b. How many Earth radii R_e in one solar radius $R_\odot$?
   c. Solar neutrinos created in the Sun's core travel at very nearly the speed of light but hardly interact with solar matter. How long does it take for such core neutrinos to reach the solar surface? How long to reach us on Earth?
   d. What then is the solar radius in light seconds? An astronomical unit in light minutes?
3. *Moon*
   The Moon is about 240 000 miles from the Earth.
   a. What is the Earth–Moon distance in kilometers? In light seconds? In Earth radii R_e? In solar radii $R_\odot$?
   b. How many Earth–Moon distances in 1 au?
4. *Ranking of sizes*
   In orders of magnitude, how much bigger is:
   a. Sun versus Earth?
   b. Galaxy versus Sun?

      c. Galaxy versus Earth?

      d. Galaxy versus us?

      e. Light year versus us?

      f. Light year versus Sun?

5. *Ranking of times*

   In orders of magnitude, how many:

      a. Human lifetimes in the age of the universe?

      b. Heartbeats in the age of the universe?

      c. Days in the age of the universe?

      d. Human lifetimes since the dinosaurs?

      e. Heartbeats since the dinosaurs?

      f. Days since the dinosaurs?

6. *Ranking of speeds*

   In orders of magnitude, how must faster is:

      a. Airplane versus slow walk?

      b. Earth's orbit versus airplane?

      c. Sun's orbit versus bicycle?

      d. Light versus car?

      e. Supernova versus Moon?

      f. Apollo versus bicycle?

2 Astronomical Distances

2.1 Angular Size

To understand ways we might infer stellar distances, let us first consider how we intuitively estimate distance in our everyday world. Two common ways are through apparent *angular size*, and/or using our *stereoscopic vision*.

For the first, let us suppose we have some independent knowledge of the physical size of a viewed object. The apparent angular size that object subtends in our overall field of view is then used intuitively by our brains to infer the object's distance, based on our extensive experience that a greater distance makes the object subtend a smaller angle.

As illustrated in Figure 2.1, we can, with the help of some elementary geometry, formalize this intuition to write the specific formula. The triangle illustrates the angle α subtended by an object of size s from a distance d. From simple trigonometry, we find

$$\tan(\alpha/2) = \frac{s/2}{d}. \tag{2.1}$$

For distances much larger than the size, $d \gg s$, the angle is small, $\alpha \ll 1$, for which the tangent function can be approximated (e.g., by first-order Taylor expansion; see Figure 2.2) to give $\tan(\alpha/2) \approx \alpha/2$, where α is measured here in radians. (1 rad = 180 degree/$\pi \approx 57$ degrees). The relation between distance, size, and angle thus becomes simply

$$\boxed{\alpha \approx \frac{s}{d}.} \tag{2.2}$$

Of course, if we know the physical size and then measure the angular size, we can solve the above relation to determine the distance $d = s/\alpha$.

As illustrated in Figure 2.3, for a spherical object the angular size α is related to the distance d and radius R through the sine function,

$$\sin(\alpha/2) = R/d. \tag{2.3}$$

From Figure 2.2 we see that, for the small angles that apply at large distances $d \gg R$, this again reduces to a simple linear form, $\alpha \approx 2R/d$, that relates size to distance.

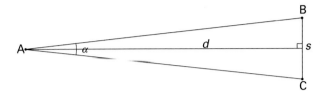

Figure 2.1 Angular size and parallax: The triangle illustrates how an object of physical size s (BC) subtends an angular size α when viewed from a point A that is at a distance d. Note that the same triangle can also illustrate the *parallax* angle α toward the point A at distance d when viewed from two points B and C separated by a length s.

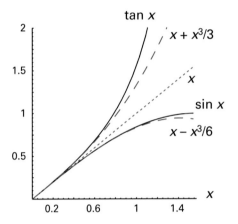

Figure 2.2 Taylor expansion of trigonometry functions $\sin x$ and $\tan x$, about $x = 0$ to order x and order x^3.

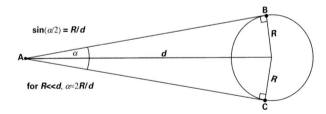

Figure 2.3 Diagram to illustrate the relation between angular size α and diameter $2R$ for a sphere at distance d .

For example, the distance from the Earth to the Sun, known as an "astronomical unit" (abbreviated "au"), is $d = 1\,\mathrm{au} \approx 150 \times 10^6$ km, much larger than the Sun's physical size (i.e., diameter), which is about $s = 2R_\odot \approx 1.4 \times 10^6$ km. Thus the distance to the Sun is a little more than 100 times its diameter, which means that it has an apparent angular diameter,

$$\alpha_\odot \approx \frac{2R_\odot}{1\mathrm{au}} \approx 0.009\,\mathrm{rad} \approx 0.5\,\mathrm{degree} = 30\,\mathrm{arcmin} = 1800\,\mathrm{arcsec}. \tag{2.4}$$

However, as noted in Section 1.2 (and illustrated in Figure 1.1), even the nearest stars are more than 200 000 times further away than the Sun. If we assume a similar physical radius (which actually is true for one of the components of the nearest star system, Alpha Centauri A), then

$$\alpha_* = \frac{2R_\odot}{200\,000\,\text{au}} \approx 0.009\,\text{arcsec}. \tag{2.5}$$

For ground-based telescopes, the distorting effect of the Earth's atmosphere, known as "atmospheric seeing" (see Section 13.2), blurs images over an angle size of about 1 arcsec, making it very difficult to infer the actual angular size. There are some specialized techniques, e.g., "speckle interferometry," that can just barely resolve the angular diameter of a few nearby giant stars (e.g., Betelgeuse, also known as Alpha Ori). But, generally, the difficulty of measuring a star's angular size means that, even if we knew its physical size, we cannot use this angular-size method to infer its distance.

2.2   Trigonometric Parallax

Fortunately, there is a practical, quite direct, way to infer distances to at least relatively nearby stars, namely through the method of *trigonometric parallax*.

This is physically quite analogous to the stereoscopic vision by which we use our two eyes to infer distances to objects in our everyday world. To understand this parallax effect, we can again refer to Figure 2.1. If we now identify s as the *separation* between the eyes, then, when we view objects at some nearby distance d, the two eyes, in order to combine the separate images as one, have to point inward at an angle $\alpha = 2\arctan(s/2d)$. Neurosensors in the eye muscles that effect this inward pointing relay this inward angle to our brain, where it is processed to provide our sense of "depth" (i.e., distance) perception.

You can easily experiment with this effect by placing your finger a few inches from your face, then blinking between your left and right eye, which thus causes the image of your finger to jump back and forth by the angle $\alpha = 2\arctan(s/2d)$. The eye separation s is fixed, but as you move the finger closer and further away, the angle shift will become respectively larger and smaller.

Home Experiment To illustrate this close link between parallax and angular size, try the following experiment. In front of a wall mirror, close one eye and then extend a finger from either arm to the mirror, covering the image of your closed eye. Without moving your finger, now switch the closure to the other eye. Note that the finger has also switched to cover the other (now closed) eye, even though you didn't physically move it! Note further that this even still works as you decrease the distance from your face to the mirror. The key point here is that the "parallax" angle shift of your finger, which results from switching perspective from one eye to the other, exactly fits the apparent angular separation between your own mirror-image eyes.

Of course, for distances much more than the separation between our eyes, $d \gg s$, the angle becomes too small to perceive, and so we can only use this approach to infer distances of about, say, 10 m. But if we extend the baseline to much larger sizes s,

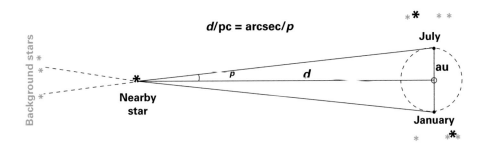

Figure 2.4 Illustration of stellar parallax, in which a relatively nearby star appears to shift against background stars by a parallax angle p as the Earth moves through the 1 au radius of Earth's orbit. The distance d in parsec (pc) is given by the inverse of p measured in arcsec.

then, when coupled with accurate measures of the angle shift α, this method can be used to infer much larger distances.

For example, in the nineteenth century, there were efforts to use this approach to infer the distance to Mars at a time when it was relatively close to Earth, namely at opposition (i.e., when Mars is on the opposite side of the Earth from the Sun). Two expeditions tried to measure the position of Mars at the same time from widely separated sites on Earth. If the distance between the sites is known, the angle difference in the measured directions to Mars, which turns out to be about 1 arcmin, yields a distance to Mars.

The largest separation possible from two points on the surface of the Earth is limited by the Earth's diameter. But to apply this method of trigonometric parallax to infer distances to stars, we need to use a much bigger baseline than the Earth's diameter. Fortunately, though, we don't need then to go into space.

As illustrated in Figure 2.4, just waiting for half a year from one place on the Earth allows us, as a result of the Earth's *orbit* around the Sun, to view the stars from two points separated by twice the Earth's orbital radius, i.e., 2 au. By convention, however, the associated "parallax angle" α of a star is traditionally quoted in terms of the shift from a baseline s of just 1 au. If we scale the parallax angle in units of an arcsec, the distance is

$$d = \frac{s}{\alpha} = \frac{206\,265\ \text{arcsec/radian}}{\alpha/\text{radian}}\ \text{au} \equiv \frac{\text{arcsec}}{\alpha}\ \text{parsec,} \qquad (2.6)$$

where we note that the conversion between arcsec and radian is given by $(180/\pi)$ degree/radian $\times$ 60 arcmin/degree $\times$ 60 arcsec/arcmin = 206 265 arcsec/radian. In the last equality, we have also introduced the distance unit *parsec* (short for "parallax second," and often further abbreviated as "pc"), which is defined as the distance at which the parallax angle is 1 arcsec. It is thus apparent that 1 pc = 206 265 au, which gives 1 pc $\approx 3 \times 10^{16}$ m.

The "parsec" is one of the two most common units used to characterize the huge distances we encounter in astronomy. The other is the *light year*, which is the distance light travels in a year, at the speed of light $c = 3 \times 10^8$ m/s. The number of seconds in a year is given by $1 \text{ yr} = 365 \times 24 \times 60 \times 60 = 3.15 \times 10^7$ s, which, coincidentally, can be remembered as $1 \text{ yr} \approx \pi \times 10^7$ s (or since $\sqrt{10} \approx 3.16$, $1 \text{ yr} \approx 10^{7.5}$ s). Thus a light year (ly) is roughly $1 \text{ ly} \approx 3\pi \times 10^{8+7} \approx 9.5 \times 10^{15} \approx 10^{16}$ m. In terms of parsecs, we can see that $1 \text{ pc} \approx 3.26 \text{ ly}$.

The parallax for even the nearest star is less than 1 arcsec, implying stars are all at distances of more (generally *much* more) than 1 parsec. By repeated observation, the roughly 1 arcsec overall blurring of single stellar images by atmospheric seeing can be averaged to give a position accuracy of about $\Delta \alpha \approx 0.01$ arcsec, implying that one can estimate distances to stars out to about $d \approx 100$ pc. The Hipparcos satellite orbiting above the Earth's atmosphere can measure parallax angles approaching 1 mas ($= 10^{-3}$ arcsec), thus potentially extending distance measurements for stars out to about 1 kiloparsec, $d \approx 1$ kpc. However, parallax measurements out to such distances typically require a relatively bright source. In practice, only a fraction of all the stars (those with the highest intrinsic brightness, or "luminosity") with distances near $d \approx 1$ kpc have thus far had accurate measurements of their parallax.[1]

Again, from the above discussion it should be apparent that parallax is really the "flip slide" of the angular size versus distance relation. That is, the triangle in Figure 2.1 was initially used to illustrate how, from the perspective of a given point A, the angle α subtended by an object is set by the ratio of its size s to its distance d. But if we consider a simple change of the observer's perspective to the two *endpoints* (B and C) of the size seqment s, then the same triangle can be used equally well to illustrate the observed parallax angle α for the point A at a distance d.

For the large ($>$pc) distances in astronomy, it is convenient to rewrite our simple (Eq. (2.2)) to scale angular size in arcsec, with the size in au and distance in pc:

$$\boxed{\frac{\alpha}{\text{arcsec}} = \frac{s/\text{au}}{d/\text{pc}}.} \qquad (2.7)$$

2.3   Determining the Astronomical Unit (au)

We thus see that determining the distance of the Earth to the Sun, i.e., measuring the physical length of an astronomical unit (au), provides a fundamental basis for determining the distances to stars and other objects in the universe. In modern times, one way this is computed involves first measuring the distance from the Earth to the planet Venus through "radar ranging," i.e., measuring the time Δt it takes a radar signal to bounce off Venus and return to Earth. The associated Earth–Venus distance is then given by

$$d_{ev} = \frac{c\Delta t}{2}. \qquad (2.8)$$

---

[1] Since 2013, a follow-up satellite mission called Gaia has been in the process of measuring the absolute position and parallax to more than a *billion* stars; see http://sci.esa.int/gaia/.

If this distance is measured at the time when Venus has its "maximum elongation," or maximum angular separation from the Sun, which is found to be about 47 degrees, then one can use simple trigonometry to derive a physical value of the astronomical unit. The details are left as an exercise for the reader; see Exercise 2.3.

2.4   Solid Angle

In general, objects that have a measurable angular size on the sky are extended in *two* independent directions. As the two-dimensional (2D) generalization of an angle along just one direction, it is useful then to define for such objects a 2D *solid angle* Ω, measured now in *square radians*, but more commonly referred by the shorthand *steradians*.

Just as projected area A is related to the square of physical size s (or radius R), so is solid angle Ω related to the square of the *angular size* α. For an object at a distance d with projected area A, the solid angle is just

$$\Omega = \frac{A}{d^2} \approx \frac{\pi R^2}{d^2} = \pi \alpha^2, \tag{2.9}$$

where the latter equalities assume a sphere (or disk) with projected radius R and associated angular radius $\alpha = R/d$.

For more general shapes, Figure 2.5 illustrates how a small solid-angle patch $\delta\Omega$ is defined in terms of ranges in the standard spherical angles representing colatitude θ and azimuth ϕ on a sphere. An extended object would then have a solid angle given by the integral

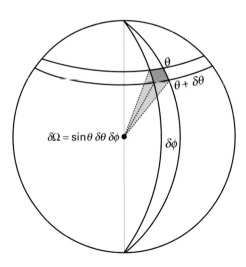

Figure 2.5 Diagram to illustrate a small patch of solid angle $\delta\Omega$ seen by an observer at the center of a sphere, with size defined by ranges in the colatitude θ and azimuth ϕ.

$$\Omega = \int d\phi \, \sin\theta \, d\theta. \tag{2.10}$$

Integration over a full sphere shows that there are 4π steradians in the full sky. This represents the 2D analog to the 2π radians around the full circumference of a circle.

For our example of a circular patch of angular radius α, let us assume the object is centered around the coordinate pole – representing perhaps the image of a distant spherical object like the Sun or Moon. The azimuthal symmetry means the ϕ integral evaluates to 2π, while carrying out the remaining integral over colatitude range 0 to α then gives

$$\Omega = 2\pi \left[1 - \cos\alpha\right]. \tag{2.11}$$

In particular, applying the angular radius of the Sun $\alpha_\odot \approx R_\odot/\text{au}$ and expanding the cosine to first order (i.e., $\cos x \approx 1 - x^2/2$), we find

$$\Omega_\odot = 2\pi \left[1 - \cos(R_\odot/\text{au})\right] \approx \pi(R_\odot/\text{au})^2 \approx \pi\alpha_\odot^2. \tag{2.12}$$

One can alternatively measure a solid angle in terms of square degrees. Since there are $180/\pi \approx 57.3$ degrees in a radian, there are $(180/\pi)^2 = 57.3^2 \approx 3283$ square degrees in a steradian (abbreviated as sr); the number of square degrees in the 4π steradians of the full sky is thus

$$4\pi \left(\frac{180}{\pi}\right)^2 = 41\,253 \, \text{degree}^2. \tag{2.13}$$

The Sun and Moon both have angular radii of about 0.25 degree, meaning they each have a solid angle of about $\pi(0.25)^2 = \pi/16 = 0.2 \, \text{degree}^2 = 6 \times 10^{-5}$ sr, which is about $1/200\,000$ of the full sky.[2]

2.5 Questions and Exercises

Quick Questions

1. Betelgeuse has a diameter of about 1800 $R_\odot$, and a distance of $d \approx 220$ pc. What is its angular diameter in milliarcseconds (mas)?
2. What is the distance (in pc) to a star with a parallax of 0.1 arcsec?
3. The average separation between human eyes is $s \approx 60$ mm. Derive a general formula for the distance d (in m) to an object with visual parallax angle α.
4. If we lived on Mars instead of Earth, what would be the length of a parsec (in km)?
5. Over a period of several years, two stars appear to go around each other with a fixed angular separation of 1 arcsec. What is the physical separation, in au, between the stars if they have a distance $d = 10$ pc from Earth?

---

[2] If you think about it, you'll see that this helps to explain why a full Moon is about a million times dimmer than full sunlight! See Exercise 3 in Section 3.4.

6. What angle α would the Earth–Sun separation subtend if viewed from a distance of $d = 1$ pc? Give your answer in both radian and arcsec. How about from a distance of $d = 1$ kpc?

Exercises

1. A helium party balloon of diameter 20 cm floats 1 m above your head.
   a. What is its angular diameter, in degrees and radians?
   b. What is its solid angle, in square degrees and steradians?
   c. What fraction of the full sky does it cover?
   d. At what height h would its angular diameter equal that of the Moon and Sun?

2. *Parallax of Mars*
   In 1672, an international effort was made to measure the parallax angle of Mars at opposition, when it was on the opposite side of the Earth from the Sun, and thus relatively close to Earth.
   a. Consider two observers at the same longitude, but one at latitude of 45 degrees North and the other at 45 degrees South. Work out the physical separation s between the observers, given the radius of Earth is $R_e \approx 6400$ km.
   b. If the parallax angle measured is 22 arcsec, what is the distance to Mars? Give your answer in both km and au.

3. *Radar to Venus*
   At the time when Venus exhibits its maximum elongation angle of about 47 degrees from the Sun, a radar signal reflected by Venus is found to take a round-trip time $\Delta t = 667$ s to return to Earth. Assuming both Earth and Venus have circular orbits, and using the speed of light $c = 3 \times 10^5$ km/s, compute (in km) the Earth–Sun distance, 1 au.

4. *Parallax extension*
   With a sufficiently large telescope in space, with angle error $\Delta\alpha \approx 1$ mas, for how many more stars can we expect to obtain a measured parallax than we can from ground-based surveys with $\Delta\alpha \approx 20$ mas? (Hint: What assumption do you need to make about the space density of stars in the region of the galaxy within 1 kpc from the Sun/Earth?)

3 Stellar Luminosity

3.1 "Standard Candle" Methods for Distance

In our everyday experience, there is another way we sometimes infer distance, namely by the change in apparent brightness for objects that emit their own light, with some known power or "luminosity." For example, a hundred-watt (100 W) light bulb at a distance of $d = 1$ m certainly appears a lot brighter than that same bulb at $d = 100$ m. Similarly, for a star, what we observe as apparent brightness is really a measure of the *flux* of light, i.e., energy per unit time *per unit area* (erg/s/cm^2 in CGS units, or W/m^2 in MKS).

When viewing a light bulb with our eyes, it is just the rate at which the light's energy is captured by the area of our pupils. If we assume the light bulb's emission is *isotropic* (i.e., the same in all directions), then as the light travels outward to a distance d, its power or luminosity is spread over a sphere of area $4\pi d^2$. This means that the light detected over a fixed detector area (like the pupil of our eye, or, for telescopes observing stars, the area of the telescope mirror) decreases in proportion to the *inverse-square* of the distance, $1/d^2$. We can thus define the apparent brightness in terms of the flux,

$$F = \frac{L}{4\pi d^2}. \tag{3.1}$$

This is a profoundly important equation in astronomy, and so you should not just memorize it, but embed it completely and deeply into your psyche.

In particular, it should become obvious that this equation can be readily used to infer the distance to an object of *known luminosity*, an approach called the *standard candle* method. (Taken from the idea that a candle, or at least a "standard" candle, has a known luminosity or intrinsic brightness.) As discussed further in the following chapters, there are circumstances in which we can get clues to a star's (or other object's) intrinsic luminosity L, for example through careful study of a star's spectrum. If we then measure the apparent brightness (i.e., flux F), we can infer the distance through:

$$d = \sqrt{\frac{L}{4\pi F}}. \tag{3.2}$$

Indeed, when the study of a stellar spectrum is the way we infer the luminosity, this method of distance determination is sometimes called "spectroscopic parallax."

Of course, if we can independently determine the distance through the actual trigonometric parallax, then such a simple measurement of the flux can instead be used to determine the luminosity,

$$L = 4\pi d^2 F. \tag{3.3}$$

In the case of the Sun, the flux measured at the Earth is referred to as the "solar constant," with a measured mean value of

$$F_\odot \approx 1.4 \, \frac{\text{kW}}{\text{m}^2} = 1.4 \times 10^6 \frac{\text{erg}}{\text{cm}^2 \, \text{s}}. \tag{3.4}$$

If we then apply the known mean distance of the Earth to the Sun, $d = 1$ au, we obtain for the solar luminosity,

$$L_\odot \approx 4 \times 10^{26} \, \text{W} = 4 \times 10^{33} \, \frac{\text{erg}}{\text{s}}. \tag{3.5}$$

Thus we see that the Sun emits the power of about 4×10^{24} 100 W light bulbs! In common language this corresponds to four million billion billion, a number so huge that it loses any meaning. It illustrates again how, in astronomy, we have to think on a entirely different scale than we are used to in our everyday world.

But once we get used to the idea that the luminosity and other properties of the Sun are huge but still finite and measurable, we can use these as benchmarks for characterizing analogous properties of other stars and astronomical objects. Stellar luminosities, for example, typically range from about $L_\odot/1000$ for very cool, low-mass "dwarf" stars, to as high as $10^6 L_\odot$ for very hot, high-mass "supergiants."

As discussed further below, the luminosity of a star depends directly on both its size (i.e., radius) and surface temperature (see Chapter 5). But, more fundamentally, these in turn are largely set by the star's mass (Section 10.4), as well as its age and evolutionary status (Section 19.1).

3.2 Intensity or Surface Brightness

For any object with a resolved solid angle Ω, an important flux-related quantity is the *surface brightness* – also known as the specific intensity I; this can be roughly (though not quite exactly; see Section 12.1) thought of as the *flux per solid angle*, i.e.,

$$I \approx \frac{F}{\Omega} \approx \frac{L}{4\pi d^2 \pi (R/d)^2} \approx \frac{L}{4\pi^2 R^2} = \frac{F_*}{\pi}, \tag{3.6}$$

where $F_* \equiv F(R) = L/4\pi R^2$ is the *surface flux* evaluated at the stellar radius R. As illustrated in Figure 3.1, the surface brightness of any resolved radiating object turns out, somewhat surprisingly, to be *independent of distance*. This is because, even though the flux declines with distance, the surface brightness "crowds" this flux into

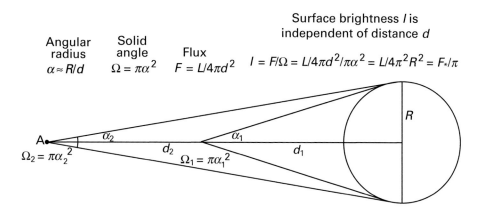

Figure 3.1 Distance independence of surface brightness of a radiating sphere, representing the flux per solid angle, $B = F/\Omega$. At greater distance d, the flux declines in proportion to $1/d^2$; but because this flux is squeezed into a smaller solid angle Ω, which also declines as $1/d^2$, the surface brightness B remains constant, independent of the distance.

a proportionally smaller solid angle as the distance is increased. The ratio of flux per solid angle, or surface brightness, is thus constant.

In particular, if we ignore any absorption from Earth's atmosphere, the surface brightness of the Sun that we see here on Earth is actually the *same* as if we were standing on the surface of the Sun itself!

Of course, on the surface of the Sun, its radiation will fill up half the sky – i.e., 2π sr, instead of the mere 0.2 degree^2 = 6×10^{-5} sr seen from Earth. The huge flux from this large, bright solid angle would cause a lot more than a mere sunburn![1]

3.3   Apparent and Absolute Magnitude and the Distance Modulus

To summarize, we have now identified three distinct kinds of "brightness" – absolute, apparent, and surface – associated respectively with the luminosity (energy/time), flux (energy/time/area), and specific intensity (flux emitted into a given solid angle). Before moving on to examine additional properties of stellar radiation, let us first discuss some specifics of how astronomers characterize apparent versus absolute brightness, namely through the so-called magnitude system.

This system has some rather awkward conventions, developed through its long history, dating back to the ancient Greeks. As noted in Chapter 1, they ranked the apparent brightness of stars in six bins of magnitude, ranging from $m = 1$ for the brightest to $m = 6$ for the dimmest. Because the human eye is adapted to detect

---

[1] In 2018, NASA launched "Parker Solar Probe," which will eventually fly within about $9R_\odot$ of the solar surface, or about $\sim 1/20$ au. So a key challenge has been to provide the shielding to keep the factor > 400 higher solar-radiation flux from frying the spacecraft's instruments.

a large dynamic range in brightness, it turns out that our perception of brightness depends roughly on the *logarithm* of the flux.

In our modern calibration this can be related to the Greek magnitude system by stating that a *difference of 5* in magnitude represents a *factor 100* in the relative brightness of the compared stars, with the *dimmer* star having the *larger magnitude*. This can be expressed in mathematical form as

$$m_2 - m_1 = 2.5 \log(F_1/F_2). \tag{3.7}$$

We can further extend this logarithmic magnitude system to characterize the absolute brightness, or luminosity, of a star in terms of an *absolute* magnitude. To remove the inherent dependence on distance in the flux F, and thus in the apparent magnitude m, the absolute magnitude M is defined as the apparent magnitude that a star *would* have if it were placed at a standard distance, chosen by convention to be $d = 10\,\mathrm{pc}$. Since the flux scales with the inverse-square of distance, $F \sim 1/d^2$, the difference between apparent magnitude m and absolute magnitude M is given by

$$m - M = 5\log(d/10\,\mathrm{pc}), \tag{3.8}$$

which is known as the *distance modulus*.

The absolute magnitude of the Sun is $M \approx +4.8$ (although for simplicity in calculations, this is often rounded up to 5), and so the scaling for other stars can be written as

$$M = 4.8 - 2.5 \log(L/L_\odot). \tag{3.9}$$

Combining these relations, we see that the apparent magnitude of any star is given in terms of the luminosity and distance by

$$m = 4.8 - 2.5 \log(L/L_\odot) + 5 \log(d/10\,\mathrm{pc}). \tag{3.10}$$

For bright stars, magnitudes can even become negative. For example, the (apparently) brightest star in the night sky, Sirius, has an apparent magnitude $m = -1.42$. But with a luminosity of just $L \approx 23L_\odot$, its absolute magnitude is still positive, $M = +1.40$. Its distance modulus, $m - M = -1.42 - 1.40 = -2.82$, is negative. Through Eq. (3.8), this implies that its distance, $d = 10^{1-2.82/5} = 2.7$ pc, is *less* than the standard distance of 10 pc used to define absolute magnitude and distance modulus in Eq. (3.8).

3.4   Questions and Exercises

Quick Questions

1. Compute $L_\odot$ given $F_\odot = 1.4\,\mathrm{W/m}^2$ and the known value of an astronomical unit (au).

2. Recalling the relationship between an au and a parsec from Eq. (2.6), use Eqs. (3.8) and (3.9) to compute the apparent magnitude of the Sun. What is the Sun's distance modulus?

3. Two stars have apparent magnitude $m_1 = +1$ and $m_2 = -1$. What is the ratio of their fluxes f_1/f_2?

4. Two stars have apparent magnitude $m_1 = +1$ and $m_2 = +6$. What is the ratio of their fluxes f_1/f_2?

Exercises

1. Suppose two stars have a luminosity ratio $L_2/L_1 = 100$.
   a. At what distance ratio d_2/d_1 would the stars have the same apparent brightness, $F_2 = F_1$?
   b. For this distance ratio, what is the difference in their apparent magnitude, $m_2 - m_1$?
   c. What is the difference in their absolute magnitude, $M_2 - M_1$?
   d. What is the difference in their distance modulus?

2. A white-dwarf supernova with peak luminosity $L \approx 10^{10} L_\odot$ is observed to have an apparent magnitude of $m = +20$ at this peak.
   a. What is its absolute magnitude M?
   b. What is its distance d (in pc and ly).
   c. How long ago did this supernova explode (in Myr)? (For simplicity of computation, you may take the absolute magnitude of the Sun to be $M_\odot \approx +5$.)

3. *Moon solid angle and albedo*
   a. Assuming the Moon reflects a fraction a (dubbed the "albedo") of sunlight hitting it, derive an expression for the ratio of apparent brightness $(F_{moon}/F_\odot)$ between the full Moon and the Sun, in terms of the Moon's radius R_{moon} and its distance from Earth, $d_{em} \ll$ au.
   b. Derive the value of the albedo a for which this ratio equals the fraction of sky subtended by the Moon's solid angle, i.e., for which $F_{moon}/F_\odot = \Omega_{moon}/4\pi$.
   c. Estimate the percentage correction to parts a and b if one accounts for the finite size of the ratio $d_{em}/$au. Be sure to make clear if it makes the answers bigger or smaller by that percentage. For the revised part b, comment on whether such an albedo a is even possible.

4. *Surface brightness of Sun and Moon*
   a. From the Sun's radius $R_\odot$ and distance of 1 au, compute the Sun's angular diameter in both degrees and radians.
   b. Use these values to compute the Sun's solid angle at the Earth, $\Omega_\odot$, in both square degrees and steradians (sr).
   c. Next compute the Sun's flux at the Earth, F_e, in W/m^2.
   d. Now compute the Sun's surface brightness $I_\odot$ in W/m^2/sr.
   e. Finally, assuming the Moon reflects a fraction $a = 0.1$ of the light hitting it, estimate its surface brightness, I_{moon}, again in W/m^2/sr. (Hint: Recall that the Moon has roughly the same angular size as the Sun.)

5. *Energy flux and magnitude*

   a. Suppose two objects have energy fluxes of f and $f + \Delta f$, where $\Delta f \ll f$. Derive an approximate expression for the magnitude difference Δm between these two objects that depends only on the *ratio* $\Delta f / f$. (Hint: Note that $\ln(1 + x) \approx x$ for $x \ll 1$.)

   b. What magnitude difference do you get for a flux difference of 10 percent.

6. *Angles, magnitudes, inverse-square law*

   a. How far from the Earth would the Sun have to be moved so that its apparent angular diameter would be 1 arcsec? (Express your answer in au.)

   b. How far away (in kilometers) would a Frisbee of diameter 30 cm have to be to subtend the same angle?

   c. At the distance you calculated in a, by what factor would the solar flux at Earth be reduced?

   d. What would the Sun's apparent magnitude be? (Use $m_\odot = -26.7$ for the actual Sun, the one that is at 1 au.)

7. *Galaxies: distance, magnitude, and solid angle*

   a. What is the apparent magnitude of a galaxy that contains 10^{11} stars identical to the Sun (i.e., assume its luminosity is equal to $10^{11} \, L_\odot$) if it is at a distance of 10 million parsecs?

   b. If the galaxy is circular in shape, as seen from the Earth, and has a diameter of $50\,000$ pc, what is its apparent angular diameter, in both radians and degrees?

   c. What solid angle does it subtend (in steradians and in square degrees)?

   d. How does the galaxy's surface brightness (energy/time/area/solid angle) compare to those of the Sun and Moon (express these as ratios)?

4 Surface Temperature from a Star's Color

Let us next consider *why* stars shine with such extreme brightness. Over the long term (i.e., millions of years), the enormous energy emitted comes from the energy generated (by nuclear fusion) in the stellar core, as will be discussed further in Chapter 18. But the more immediate reason for why stars shine is more direct, namely because their surfaces are so very *hot*. The light they emit is called "thermal radiation," and arises from the jostling of the atoms (and particularly the electrons in and around those atoms) by the violent collisions associated with the star's high temperature.[1]

4.1 The Wave Nature of Light

To lay the groundwork for a general understanding of the key physical laws governing such thermal radiation and how it depends on temperature, we have to review what is understood about the basic nature of light, and the processes by which it is emitted and absorbed.

The nineteenth-century physicist James Clerk Maxwell developed a set of four equations (Maxwell's equations) that showed how variations in electric and magnetic fields could lead to oscillating wave solutions, which he indeed indentifed with light, or, more generally, with *electromagnetic (EM) radiation*. The wavelengths λ of these EM waves are kcy to their properties. As illustrated in Figure 4.1, visible light corresponds to wavelengths ranging from $\lambda \approx 400$ nm (violet) to $\lambda \approx 750$ nm (red), but the full spectrum extends much further, including ultraviolet (UV), X-rays, and gamma rays at shorter wavelengths, and infrared (IR), microwaves, and radio waves at longer wavelengths. White light is made up of a broad mix of visible light ranging from red through green to blue (RGB).

In a vacuum, all these EM waves travel at the *same speed*, namely the speed of light, customarily denoted as c, with a value $c \approx 3 \times 10^5$ km/s $= 3 \times 10^8$ m/s $= 3 \times 10^{10}$ cm/s. The wave *period* is the time it takes for a complete wavelength to pass a fixed point at this speed, and so is given by $P = \lambda/c$. We can thus see that the sequence of

---

[1] In astronomy, temperature is measured in a degree unit called a kelvin, abbreviated K, and defined relative to the centigrade or Celsius scale $°C$ such that $K = °C + 273$. A temperature of $T = 0$ K is called "absolute zero," and represents the ideal limit where all thermal motion is completely stopped. To convert from the US use of the Fahrenheit scale F, we first just convert to centigrade using $°C = (5/9)(F - 32)$, and then add 273 to get the temperature in kelvin (K).

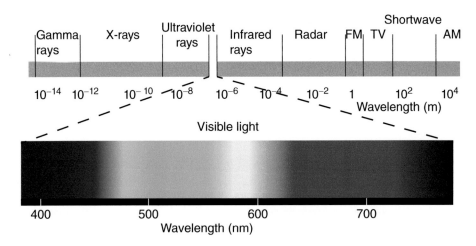

Figure 4.1 The electromagnetic spectrum.

wave crests passes by at a *frequency* of once per period, $\nu = 1/P$, implying a simple relationship between light's wavelength λ, frequency ν, and speed c,

$$\boxed{\frac{\lambda}{P} = \lambda\nu = c.}$$ (4.1)

4.2   Light Quanta and the Blackbody Emission Spectrum

The wave nature of light has been confirmed by a wide range of experiments, for example through the diffraction and interference when light passes by structures with a scale comparable to the wavelength. Indeed, a diffraction grating of narrowly etched lines on a glass plate is a principal way by which astronomers spread light into a spatially separated spectrum, much as the spread that occurs by refraction through a prism.

However, at the beginning of the twentieth century, work by Einstein, Planck, and others led to the realization that light waves are also *quantized* into discrete wave "bundles" called *photons*. Each photon carries a discrete, indivisible "quantum" of energy that depends on the wave frequency as

$$\boxed{E = h\nu,}$$ (4.2)

where h is *Planck's constant*, with value $h \approx 6.6 \times 10^{-27}$ erg s $= 6.6 \times 10^{-34}$ J s.

This quantization of light (and indeed of all energy) has profound and wide-ranging consequences, most notably in the current context for how thermally emitted radiation is distributed in wavelength or frequency. This is known as the "spectral energy distribution" (SED). For a so-called *blackbody* – meaning idealized material that is readily able to absorb and emit radiation of all wavelengths – Planck showed that, as thermal motions of the material approach a *thermodynamic equilibrium* (TE) in the exchange of energy between radiation and matter, the SED can be described by a function that depends *only* on the gas *temperature T* (and *not*, e.g., on the density, pressure, or chemical composition).

In terms of the wave frequency ν, this *Planck blackbody* function takes the form

$$B_\nu(T) = \frac{2h\nu^3/c^2}{e^{h\nu/kT} - 1},$$

(4.3)

where k is Boltzmann's constant, with value $k = 1.38 \times 10^{-16}$ erg/K $= 1.38 \times 10^{-23}$ J/K. For an interval of frequency between ν and $\nu + d\nu$, the quantity $B_\nu d\nu$ gives the emitted energy per unit time per unit area *per unit solid angle*. This means the Planck blackbody function is fundamentally a measure of *intensity* or *surface brightness*, with B_ν representing the *distribution* of surface brightness over frequency ν, having CGS units erg/cm^2/s/sr/Hz (and MKS units W/m^2/sr/Hz).

Sometimes it is convenient instead to define this Planck distribution in terms of the brightness distribution in a *wavelength* interval between λ and $\lambda + d\lambda$, $B_\lambda d\lambda$. Requiring that this equals $B_\nu d\nu$, and noting that $\nu = c/\lambda$ implies $|d\nu/d\lambda| = c/\lambda^2$, we can use Eq. (4.3) to obtain

$$B_\lambda(T) = \frac{2hc^2/\lambda^5}{e^{hc/\lambda kT} - 1}.$$

(4.4)

4.3 Inverse-Temperature Dependence of Wavelength for Peak Flux

Figure 4.2 plots the variation of B_λ versus wavelength λ for various temperatures T. Note that, for higher temperature, the level of B_λ is higher at *all* wavelengths, with greatest increases near the peak level.

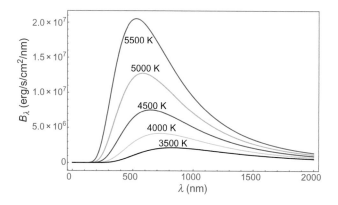

Figure 4.2 The Planck blackbody spectral energy distribution (SED) versus wavelength λ, plotted for various temperatures T.

Moreover, the location of this peak shifts to *shorter* wavelength with *higher* temperature. We can determine this peak wavelength λ_{max} by solving the equation

$$\left[\frac{dB_\lambda}{d\lambda}\right]_{\lambda=\lambda_{max}} \equiv 0. \tag{4.5}$$

Leaving the details as an exercise, the result is

$$\boxed{\lambda_{max} = \frac{2.9 \times 10^6 \text{ nm K}}{T} = \frac{290 \text{ nm}}{T/10\,000 \text{ K}} \approx \frac{500 \text{ nm}}{T/T_\odot},} \tag{4.6}$$

which is known as *Wien's displacement law*.

For example, the last equality uses the fact that the observed wavelength peak in the Sun's spectrum is $\lambda_{max,\odot} \approx 500$ nm, very near the the middle of the visible spectrum.[2] Using this, one can solve for a blackbody-peak estimate for the Sun's surface temperature

$$T_\odot = \frac{2.9 \times 10^6 \text{ nm K}}{500 \text{ nm}} = 5800 \text{ K}. \tag{4.7}$$

By similarly measuring the peak wavelength λ_{max} in other stars, we can likewise derive an estimate of their surface temperature by

$$\boxed{T = T_\odot \frac{\lambda_{max,\odot}}{\lambda_{max}} \approx 5800 \text{ K} \frac{500 \text{ nm}}{\lambda_{max}}.} \tag{4.8}$$

4.4   Inferring Stellar Temperatures from Photometric Colors

In practice, this is not quite the approach to estimating a star's temperature that is most commonly used in astronomy, in part because, with real SEDs, it is relatively difficult to identify accurately the peak wavelength. Moreover, in surveying a large number of stars, it requires a lot more effort (and telescope time) to measure the full SED, especially for relatively faint stars. A simpler, more common method is just to measure the stellar *color*.

But rather than using the red, green, and blue (RGB) colors we perceive with our eyes, astronomers typically define a set of standard colors that extend to wavebands beyond just the visible spectrum. The most common example is the Johnson three-color UBV (ultraviolet, blue, visual) system. The left panel of Figure 4.3 compares the wavelength sensitivity of such UBV filters to that of the human eye. By passing the star's light through a standard set of filters designed to let through only light for the defined color waveband, the observed apparent brightness in each filter can be used to define a set of color magnitudes, e.g., m_U, m_B, and m_V.

---

[2] This is not entirely coincidental, because our eyes evolved to use the wavelengths of light for which the solar illumination is brightest.

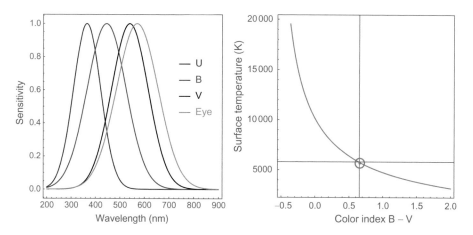

Figure 4.3 Left: Comparison of the spectral sensitivity of the human eye with those of the UBV filters in the Johnson photometric color system. Right: Temperature dependence of the B – V color for a blackbody-emitted spectrum. The circle dot marks the solar values $T_\odot \approx 5800\,\mathrm{K}$ and $(B - V)_\odot \approx 0.656$.

The standard shorthand is simply to denote these color magnitudes just by the capital letter alone, namely U, B, and V. The *difference* between two color magnitudes, e.g., $B - V \equiv m_B - m_V$, is independent of the stellar distance, but provides a direct diagnostic of the stellar temperature, sometimes called the "color temperature."

Because a larger magnitude corresponds to a lower brightness, stars with a positive B – V actually are less bright in the blue than in the visible, implying a relatively *low* temperature. On the other hand, a negative B – V means blue is brighter, implying a *high* temperature. The right panel of Figure 4.3 shows how the temperature of a blackbody varies with the B – V color of the emitted blackbody spectrum.

4.5   Questions and Exercises

Quick Questions

1. Derive Eq. (4.6) from Eq. (4.4) using the definition (4.5).
2. Using $B_\nu d\nu = B_\lambda d\lambda$ and the relationship between frequency ν and wavelength λ, derive Eq. (4.4) from Eq. (4.3).
3. As will be discussed in Chapter 32, the Cosmic Microwave Background (CMB) has a temperature of 2.7 K. What is its peak wavelength, both in nm and mm?
4. Two photons have wavelength ratio $\lambda_2/\lambda_1 = 2$.
   a. What is the ratio of their period P_2/P_1?

b. What is the ratio of their frequency ν_2/ν_1?

c. What is the ratio of their energy E_2/E_1?

5. The star Alpha Centauri A has a luminosity $L = 1.5L_\odot$, and lies at a distance of 4.37 ly. What are its absolute magnitude M and apparent magnitude m? What is its distance modulus?

6. Two stars have the same luminosity, but star 2 is 10 times as far as star 1. Give the ratio of their fluxes f_2/f_1 and the difference in the apparent magnitude $m_2 - m_1$.

7. From Figure 4.3, estimate the temperatures T (in K) of stars with colors $B - V = 1$ and $B - V = 0$.

8. Assuming the Earth has an average temperature equal to that of typical spring day, i.e., 50 °F, calculate the peak wavelength of Earth's blackbody radiation. What part of the EM spectrum does this lie in?

Exercises

1. *Stellar temperatures*
    a. Estimate the temperature of stars with $\lambda_{\max} = 100$, 300, 1000, and 3000 nm. (To simplify the numerics, you may take $T_\odot \approx 6000$ K, and $\lambda_{\max,\odot} = 500$ nm.)
    b. Conversely, estimate the peak wavelengths $\lambda_{\max}$ of stars with $T = 2000$, 10 000, and 60 000 K.
    c. What parts of the EM spectrum (i.e., UV, visible, IR) do each of these lie in?

2. *Equilibrium temperature of Earth*
    a. Assuming Earth is a blackbody, use the known luminosity and distance of the Sun to estimate Earth's average equilibrium surface temperature if the solar energy it intercepts is reradiated to space according the Stefan–Boltzman law. Compare this to the temperature on a moderate spring day.
    b. Earth has an albedo of $a = 0.3$, meaning this is the fraction of received light that is reflected by, e.g., clouds, snow, etc., without contributing any heat to the Earth. So now redo the calculation in part a, reducing the solar input energy by $1 - a$.
    c. Which result seems more "reasonable"? Briefly discuss what other physics might be important to include to understand the actual surface temperature of Earth.

3. *Peak of blackbody functions*
    a. For a given temperature T, derive an explicit expression for the frequency $\nu_{\max}$ at which the Planck blackbody spectrum $B_\nu(T)$ is maximum. (Note that you'll have to solve an equation numerically. I would recommend "fixed-point iteration" – you can check Google or maybe Wiki, but here's my summary: if you have an expression of the form $x = a + be^x$, you can guess a value of x, evaluate the right-hand side for this guessed value, use this new value as x, and repeat until your answer converges.)
    b. Convert this frequency to an associated wavelength $\lambda_{\nu\max}$.
    c. Compare this with the $\lambda_{\max}$ given in Eq. (4.6) for the wavelength at which the blackbody spectrum $B_\lambda(T)$ is maximum. Why do these two wavelengths differ?

4. *Brightness of the full Moon*
   a. Given the Sun's radius and distance to Earth, compute the Sun's solid angle on the sky, in steradians.
   b. Now convert this to a fraction of the full-sky's 4π steradians.
   c. By coincidence, the Moon covers about this same fraction. Use this fact, together with the Moon's albedo of 0.12 (meaning the Moon reflects back 12 percent of the sunlight that hits its surface), to estimate how much dimmer the full Moon appears compared with sunlight on Earth.

5   Stellar Radius from Luminosity and Temperature

5.1   Bolometric Intensity

We can see from Figure 4.2 that, in addition to a shift toward shorter peak wavelength $\lambda_{\max}$, a higher temperature also increases the overall brightness of blackbody emission at *all* wavelengths. This suggests that the total energy emitted over all wavelengths should increase quite sharply with temperature. Leaving the details as an exercise for the reader, let us quantify this expectation by carrying out the necessary spectral integrals to obtain the temperature dependence of the *bolometric*[1] intensity of a blackbody

$$B(T) \equiv \int_0^\infty B_\lambda(T)\, d\lambda = \int_0^\infty B_\nu(T)\, d\nu = \frac{\sigma_{\mathrm{sb}} T^4}{\pi}, \tag{5.1}$$

with $\sigma_{\mathrm{sb}} = 2\pi^5 k^4/(15 h^3 c^2)$ known as the Stefan–Boltzmann constant, with numerical value $\sigma_{\mathrm{sb}} = 5.67 \times 10^{-5}$ erg/cm^2/s/K^4 = 5.67×10^{-8} J/m^2/s/K^4.

If we spatially resolve a pure blackbody with surface temperature T, then $B(T)$ represents the bolometric *surface brightness* we would observe from each part of the visible surface.

5.2   The Stefan–Boltzmann Law for Surface Flux from a Blackbody

Combining Eqs. (3.6) and (5.1), we see that the radiative *flux* at the surface radius R of a blackbody is given by

$$\boxed{F_* \equiv F(R) = \pi\, B(T) = \sigma_{\mathrm{sb}} T^4,} \tag{5.2}$$

which is known as the *Stefan–Boltzmann law*.

The Stefan–Boltzmann law is one of the linchpins of stellar astronomy. If we now relate the surface flux to the stellar luminosity L over the surface area $4\pi R^2$, then applying this to the Stefan–Boltzmann law gives

$$\boxed{L = \sigma_{\mathrm{sb}} T^4\, 4\pi R^2,} \tag{5.3}$$

---

[1] "Bolometric" simply means summed over all wavelengths.

which is often more convenient to scale by associated solar values,

$$\frac{L}{L_\odot} = \left(\frac{T}{T_\odot}\right)^4 \left(\frac{R}{R_\odot}\right)^2. \tag{5.4}$$

We can also use Eq. (5.3) to solve for the stellar radius,

$$R = \sqrt{\frac{L}{4\pi\sigma_{sb}T^4}} = \sqrt{\frac{F(d)}{\sigma_{sb}T^4}}\, d, \tag{5.5}$$

where the latter equation uses the inverse-square law to relate the stellar radius to the flux $F(d)$ and distance d, along with the surface temperature T.

For a star with a known distance d, e.g., by a measured parallax, measurement of apparent magnitude gives the flux $F(d)$, while measurement of the peak wavelength λ_{max} or color (e.g., B − V) provides an estimate of the temperature T (see Figure 4.3). Applying these in Eq. (5.5), we can thus obtain an estimate of their stellar radius R.

5.3   Questions and Exercises

Quick Questions

1. What is the luminosity (in $L_\odot$) of a star with 10 times the solar temperature and 10 times the solar radius?
2. What is the radius (in $R_\odot$) of a star that is both twice as luminous and twice as hot as the Sun?
3. Suppose star 2 is both twice as hot and twice as far as star 1, but they have the same apparent magnitude. What is the ratio of their stellar radii, R_2/R_1?
4. A red-giant star has a temperature of 3000 K and luminosity $L = 1000L_\odot$. About what is its radius, R (in $R_\odot$)?

Exercises

1. Compute the luminosity L (in units of the solar luminosity $L_\odot$), absolute magnitude M, and peak wavelength λ_{max} (in nm) for stars with
   a. $T = T_\odot$; $R = 10R_\odot$,
   b. $T = 10T_\odot$; $R = R_\odot$, and
   c. $T = 10T_\odot$; $R = 10R_\odot$.
   d. If these stars all have a parallax of $p = 0.001$ arcsec, compute their associated apparent magnitudes m.
2. Suppose a star has a parallax $p = 0.01$ arcsec, peak wavelength $\lambda_{max} = 250$ nm, and apparent magnitude $m = +5$. Approximately what is its:
   a. Distance d (in pc)?
   b. Distance modulus $m - M$?
   c. Absolute magnitude M?
   d. Surface temperature T (in $T_\odot$)?
   e. Radius R (in $R_\odot$)?

      f. Angular radius α (in radian and arcsec)?

      g. Solid angle Ω (in steradian (sr) and arcsec^2)?

      h. Surface brightness relative to that of the Sun, $B/B_\odot$?

3. *Stellar parameters*

The star Dschubba (δ Sco) has a parallax $p = 8$ mas. Assuming it is a spherical blackbody with radius $R = 7.5\ R_\odot$ and surface temperature $T = 28\,000$ K, compute Dschubba's:

      a. Luminosity, in erg/s, and $L_\odot$.

      b. Absolute magnitude.

      c. Apparent magnitude.

      d. Distance modulus.

      e. Radiant flux at the star's surface, in CGS, and relative to surface flux of the Sun.

      f. Radiant flux at the Earth's surface, and the ratio of this to the solar irradiance.

      g. Peak wavelength λ_{max}.

6 Composition and Ionization from Stellar Spectra

6.1 Spectral Line Absorption and Emission

In reality, stars are not perfect blackbodies, and so their emitted spectra do not just depend on temperature, but contain detailed signatures of key physical properties such as elemental composition. The energy we see emitted from a stellar surface is generated in the very hot interior and then diffuses outward, following the strong temperature decline to the surface. The atoms and ions that absorb and emit the light do not do so with perfect efficiency at all wavelengths, which is what is meant by the "black" in "blackbody." We experience this all the time in our everyday world, wherein different objects have distinct "color," meaning they absorb certain wavebands of light, and reflect others. For example, a green leaf reflects some of the "green" parts of the visible spectrum – with wavelengths near $\lambda \approx 5100$ Å – and absorbs most of the rest.

For atoms in a gas, the ability to absorb, scatter, and emit light can likewise depend on the wavelength, sometimes quite sharply. As discussed in Section C.2 in Appendix C, and illustrated in the middle row of Figure C.1, the energy of electrons orbiting an atomic nucleus has discrete levels, much like the steps in a staircase. Absorption or scattering by the atom is thus much more efficient for those select few photons with an energy that closely matches the energy difference between two of these atomic energy levels.

The evidence for this is quite apparent if we examine carefully the actual spectrum emitted by any star. Although the overall spectral energy distribution (SED) discussed in Chapter 4 often roughly fits a Planck blackbody function, careful inspection shows that light is missing or reduced at a number of discrete wavelengths or colors. As illustrated in Figure 6.1 for the Sun, when the color spectrum of light is spread out, for example by a prism or diffraction grating, this missing light appears as a complex series of relatively dark "absorption lines."

Figure 6.2 illustrates how the absorption by relatively cool, low-density atoms in the upper layers of the Sun or a star's atmosphere can impart this pattern of absorption lines on the continuum, nearly blackbody spectrum emitted by the denser, hotter layers.

A key point here is that the discrete energies levels associated with atoms of different elements (or, as discussed in Section 6.3, different "ionization stages" of a given element) are quite distinct. As such, the associated wavelengths of the absorption lines in a star's spectrum provide a direct "fingerprint" – perhaps even more akin

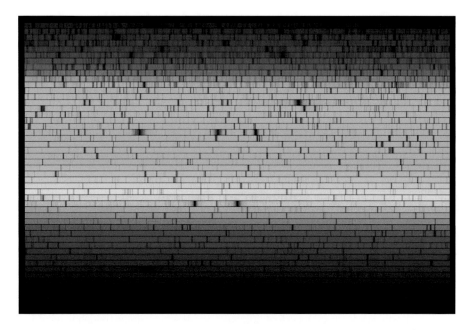

Figure 6.1 The Sun's spectrum, showing the complex pattern of absorption lines at discrete wavelength or colors. Credit: NOAO/AURA/NSF.

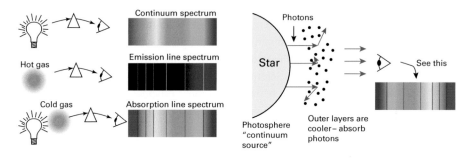

Figure 6.2 Illustration of principles for producing an emission line spectrum versus an absorption line spectrum. The left panel shows that an incandescent light passed through a prism generally produces a featureless continuum spectrum, but a cold gas placed in front of this yields an absorption line spectrum. That same gas, when heated and seen on its own against a dark background, produces the same pattern of lines, but now in *emission* instead of absorption. The right panel shows heuristically how the relatively cool gas in the surface layers of a star leads to an absorption line spectrum from the star.

to a supermarket *bar code* – for the presence of that element in the star's atmosphere. The code "key" can come from laboratory measurement of the line spectrum from known samples of atoms and ions, or, as discussed in Section A.1 in Appendix A, from theoretical models of the atomic energy levels using modern principles of quantum physics.

6.2   Elemental Composition of the Sun and Stars

With proper physical modeling, the relative strengths of the absorption lines can even provide a quantitative measure of the relative *abundance* of the various elements. A key result is that the composition of the Sun, which is typical of most all stars (and indeed of the "cosmic" abundance of elements in the present-day universe), is

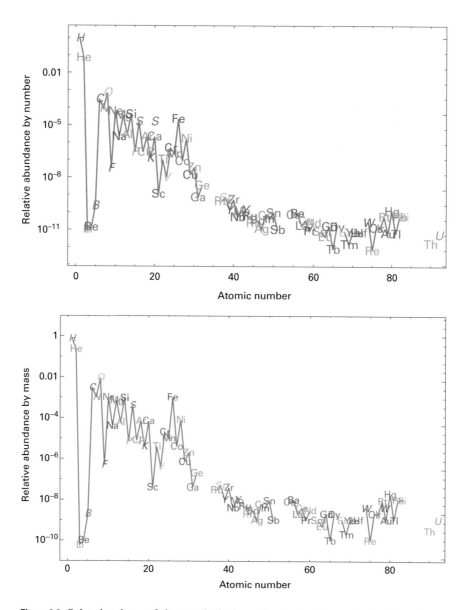

Figure 6.3 Solar abundance of elements by both *number* (top) and *mass* (bottom) fractions, plotted on a log scale versus atomic number, with data points labeled by the symbols for each element.

dominated by just the two simplest elements, namely hydrogen (H) and helium[1] (He); these make up respectively 90.9 percent and 8.9 percent of the atoms, with all the other elements numbering only 0.2 percent. The top panel of Figure 6.3 gives a log plot of these number fractions versus atomic number.

The bottom panel of Figure 6.3 shows the corresponding *mass* fractions, which now give $X \approx 0.72$ and $Y \approx 0.26$ for hydrogen and helium. All the remaining elements of the periodic table – commonly referred to in astronomy as "metals" – make up just the final 2 percent of the mass, denoted as a "metallicity" $Z \approx 0.02$. Of these, the most abundant are oxygen, carbon, and iron, with respective mass fractions of 0.009, 0.003, and 0.001.

As discussed in Chapters 18, 19, and 20, these heavier elements were all synthesized from lighter ones by nuclear fusion in the hot cores of stars, and then spewed out into space by outflows and explosions, enriching the interstellar medium so that later, in generations of stars such as the Sun, they made up about 2 percent of its mass. Then, like all the planets in our solar system, the Earth formed out of the same material that makes up the Sun (Chapter 23). But its relatively weak gravity has allowed a lot of the light elements such as hydrogen and helium to escape into space, leaving behind the heavier elements that make up our world, and us (Chapter 24). Indeed, once the hydrogen and helium are removed, the *relative* abundances of all these elements are roughly the same on the Earth as in the Sun!

6.3   Stellar Spectral Type: Ionization as a Temperature Diagnostic

Another key factor in the observed stellar spectra is that the atomic elements present are generally not electrically neutral, but typically have had one or more electrons stripped – ionized – by thermal collisions with characteristic energies set by the temperature. As discussed in Section B.2 in Appendix B, this *ionization equilibrium* depends on temperature and density. For characteristic densities in the visible surface of a star, the observed degree of ionization thus provides a diagnostic of its surface temperature.

Figure 6.4 compares the spectra of stars of different surface temperature, showing that this leads to gradual changes and shifts in the detailed pattern of absorption lines from the various ionizations stages of the various elements. The letters "OBAFGKM" represent various categories, known as spectral class or "spectral type," assigned to stars with different spectral patterns. It turns out that type O is the hottest, with temperatures about 50 000 K, while M is the coolest[2] with temperatures of about 3500 K.

---

[1] Indeed, helium was first identified as a element through early spectra of the Sun, and is even named after the Sun, through the Greek Sun god *Helios*.

[2] In recent years, it has become possible to detect even cooler "brown-dwarf" stars, with spectral classes LTY, extending down to temperatures as low as 1000 K. Brown-dwarf stars have too low a mass ($< 0.08 M_\odot$) to force hydrogen fusion in their interior; see Section 16.3. They represent a link to gas-giant planets like Jupiter, which has a mass $M_J \approx 0.001 M_\odot$.

Figure 6.4 Stellar spectra for the full range of spectral types OBAFGKM, corresponding to a range in stellar surface temperature from hot to cool. Credit: NOAO/AURA/NSF.

The sequence is often remembered through the mnemonic[3] "Oh, Be A Fine Gal/Guy Kiss Me." In keeping with its status as a kind of average star, the Sun has spectral type G, just a bit cooler than type F in the middle of the sequence.

In addition to the spectral classes OBAFGKM that depend on surface temperature T, spectra can also be organized in terms of *luminosity* classes, conventionally denoted though Roman numerals I for the biggest, brightest "supergiant" stars, to V for smaller, dimmer "dwarf" stars; in between, there are luminosity classes II (bright giants), III (giants), and IV (subgiants).

In this two-parameter scheme, the Sun is classified as a G2V star.

Finally, in addition to giving information on the temperature, chemical composition, and other conditions of a star's atmosphere, these absorption lines provide convenient "markers" in the star's spectrum. As discussed in Section 9.2, this makes it possible to track small changes in the wavelength of lines that arise from the so-called Doppler effect as a star moves toward or away from us.

In summary, the appearance of absorption lines in stellar spectra provides a real treasure trove of clues to the physical properties of stars.

6.4   Hertzsprung–Russell (H–R) Diagram

A key diagnostic of stellar populations comes from the *Hertzsprung–Russell* (H–R) diagram, illustrated by the left panel of Figure 6.5. Observationally, it relates absolute magnitude (or luminosity class) on the y-axis, to color or spectral type on the x-axis; physically, it relates luminosity to temperature. For stars in the solar neighborhood with parallaxes measured by the Gaia astrometry satellite, one can readily use the associated distance to convert observed apparent magnitudes to absolute magnitudes

---

[3] A student in one of my exams once offered an alternative mnemonic: "Oh Boy, Another F's Gonna Kill Me."

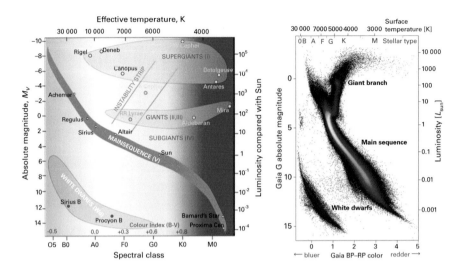

Figure 6.5 Left: Schematic Hertzsprung–Russell (H–R) diagram relating stellar absolute magnitude (or log luminosity) versus surface temperature, as characterized by the spectral type or color, with hotter bluer stars on the left, and cooler redder stars on the right. The main sequence (MS) represents stars fusing hydrogen into helium in their core, whereas the giants or supergiants are stars that have evolved away from the MS after exhausting hydrogen in their cores. The white-dwarf stars are dying remnants of solar-type stars. Image credit: CSIRO Radio Astronomy Image Archive. Right: More than 4 million stars within 5000 light years from the Sun plotted as an H–R diagram using information about their brightness, color, and distance from ESA's Gaia satellite. Credit: ESA/Gaia/DPAC, CC BY-SA 3.0 IGO.

and luminosities. The right panel of Figure 6.5 shows the H–R diagram for these stars, plotting their known luminosities versus their colors or spectral types.

The extended band of stars running from the upper left to lower right is known as the *main sequence*, representing "dwarf" stars of luminosity class V. The reason there are so many stars in this main-sequence band is that it represents the long-lived phase when stars are stably burning hydrogen into helium in their cores (Chapter 18).

The medium horizontal band above the main sequence represents "giant stars" of luminosity class III. They are typically stars that have exhausted hydrogen in their core, and are now getting energy from a combination of hydrogen burning in a shell around the core, and burning helium into carbon in the cores themselves (Chapter 19).

The relative lack here of still more luminous supergiant stars of luminosity class I stems from both the relative rarity of stars with sufficiently high mass to become this luminous, coupled with the fact that such luminous stars live for only a very short time (Section 8.4). As such, there are only a few such massive, luminous stars in the solar neighborhood. Studying them requires broader surveys extending to larger distances that encompass a greater fraction of our Galaxy.

The stars in the band below the main sequence are called *white dwarfs*; they represent the slowly cooling remnant cores of low-mass stars such as the Sun (Section 19.4).

This association between position on the H–R diagram, and stellar parameters and evolutionary status, represents a key link between the observable properties of light emitted from the stellar surface and the physical properties associated with the stellar interior. Understanding this link through examination of stellar structure and evolution will constitute the major thrust of our studies of stellar interiors in Part II of this book.

But before we can do that, we need to consider ways that we can empirically determine the two key parameters differentiating the various kinds of stars on this H–R diagram, namely *mass* and *age*.

6.5   Questions and Exercises

Quick Questions

1. On the H–R diagram, where do we find stars that are:

   a. Hot and luminous?

   b. Cool and luminous?

   c. Cool and dim?

   d. Hot and dim?

2. Of the above, which of these are known as:

   a. White dwarfs?

   b. Red giants?

   c. Blue supergiants?

   d. Red dwarfs?

3. For the Gaia H–R diagram in Figure 6.5, about what is the luminosity ratio between a white-dwarf star (at, say, color index 1), and a main-sequence star of the same color? What is the implied ratio in the stellar radii, R_{wd}/R_{ms}?

4. Referring to Figure 6.3, what is the typical factor difference between the abundance fractions of odd versus even atomic numbers? Discuss possible reasons for this difference? (Hint: Which can be made directly by fusion of helium?)

Exercises

1. Referring to Appendix B on ionization and excitation and Section A.1 in Appendix A on the Bohr model of the atom, the $n = 2$ level of hydrogen has $g_2 = 8$ states, while the ground level has just $g_1 = 2$ states. Using the energy difference ΔE_{21} from the Bohr atom, compute the Boltzmann equilibrium number ratio n_2/n_1 of electrons in these levels for a temperature $T = 100\,000$ K.

2. *Paschen series of hydrogen*
   a. Compute the wavelengths (in nm) for Paschen-α λ_{43} and the Paschen limit $\lambda_{\infty 3}$.
   b. What are the associated changes in energy (in eV), ΔE_{43} and $\Delta E_{\infty 3}$?
3. For an electron and proton that are initially a distance r apart, show that the energy needed to separate them to an arbitrarily large distance is given by e^2/r. Use the resulting potential energy $U(r)$ together with the orbital kinetic energy $T = m_e v^2/2$ to derive the expressions in Eq. (A.6) for the total energy $E = U + T$.
4. Confirm the validity of Eq. (A.6) by using Eq. (A.1) to show that $E = U/2 = -T$, where U, T, and E are the potential, kinetic, and total energy of an orbiting electron. (Note: This result is sometimes referred to as a corollary of the Virial Theorem for bound systems, which is discussed in Section 7.4.)
5. *Exponential decline in abundance fractions with atomic number*
   For the elements from C to Ti, draw by eye a best-fit straight line to the number and mass fraction plots in Figure 6.3.
   a. Use the slope of this line to derive a baseline exponential scaling formula for both the mass and fractional abundances.
   b. Specifically, for atomic number n, write these fits in the form $f(n) = A \, 10^{-bn}$.
   c. Give values for A and b for both number and mass fractions, and discuss the reasons for the differences between them.
   d. Finally, relative to this baseline, by about what factors $\mathcal{F}$ are the abundance of the elements Fe and Ni enhanced? Discuss why this might be. (Hint: See Figure 20.1.)
6. *Bohr atom for HeII: wavelength, energy, and temperature* (See Section A.1 in Appendix A.)
   Singly ionized helium (HeII) has one bound electron and thus can be analyzed using the same Bohr model used for hydrogen if one just accounts for the double positive charge from the two protons in the nucleus.
   a. Compared with the 13.6 eV needed to ionize hydrogen from its ground state $n = 1$, what factor greater energy is needed to further ionize HeII from its ground state?
   b. Use this with the Bohr model to write an expression for the binding energy E_n (in eV) of each quantum level n of HeII.
   c. Compute the energy (in eV) of the Balmer-α (between $n = 2$ and 3) transition of HeII.
   d. What is the wavelength λ of a HeII Balmer-α photon (in nm)?
   e. What temperature T has a characteristic energy kT equal to the energy in part c?

7 Surface Gravity and Escape/Orbital Speed

So far we have been able to finds ways to estimate the first five stellar parameters on our list – distance, luminosity, temperature, radius, and elemental composition. Moreover, we have done this with just a few, relatively simple measurements – parallax, apparent magnitude, color, and spectral line patterns. But along the way we have had to learn to exploit some key geometric principles and physical laws – angular-size/parallax, inverse-square law, and Planck's, Wien's, and the Stefan–Boltzmann laws of blackbody radiation.

So what of the next item on the list, namely stellar mass? Mass is clearly a physically important parameter for a star, because, for example, it will help to determine the strength of the gravity that tries to pull the star's matter together. To lay the groundwork for discussing one basic way we can determine mass (from orbits of stars in stellar binaries), let us first review Newton's law of gravitation and show how this sets key quantities such as the surface gravity, and the speeds required for material to escape or orbit the star.

7.1 Newton's Law of Gravitation and Stellar Surface Gravity

On Earth, an object of mass m has a weight given by

$$F_{grav} = mg_e, \tag{7.1}$$

where the acceleration of gravity on Earth is $g_e = 980 \, \text{cm/s}^2 = 9.8 \, \text{m/s}^2$. But this comes from Newtons's law of gravity, which states that, for two point masses m and M separated by a distance r, the attractive gravitational force between them is given by

$$\boxed{F_{grav} = \frac{GMm}{r^2},} \tag{7.2}$$

where Newton's constant of gravity is $G = 6.7 \times 10^{-8} \text{cm}^3/\text{g/s}^2$. Remarkably, when applied to spherical bodies of mass M and finite radius R, the same formula works for all distances $r \geq R$ at or outside the surface![1] Thus, we see that the acceleration

---

[1] Even more remarkably, even if we are *inside* the radius, $r < R$, then we can still use Newton's law if we just count that part of the total mass that is *inside* r, i.e., M_r, and completely ignore all the mass that is above r.

of gravity at the surface of the Earth is just given by the mass and radius of the Earth through

$$g_e = \frac{GM_e}{R_e^2}. \tag{7.3}$$

Similarly, for stars, the surface gravity is given by the stellar mass M and radius R. For the Sun, this gives $g_\odot = 2.6 \times 10^4$ cm/s$^2 \approx 27\, g_e$. Thus, if you could stand on the surface of the Sun, your "weight" would be about 27 times what it is on Earth.

For other stars, gravities can vary over a quite wide range, largely because of the wide range in size. For example, when the Sun gets near the end of its life about 5 billion years from now, it will swell up to more than 100 times its current radius, becoming what's known as a "red giant" (Chapter 19). Stars we see now that happen to be in this red-giant phase thus tend to have quite low gravity, about a fraction 1/10 000 that of the Sun.

Largely because of this very low gravity, much of the outer envelope of such red-giant stars will actually be lost to space (forming, as we shall see, quite beautiful nebulae; see Section 19.1 and Figure 20.5.) When this happens to the Sun, what is left behind will be just the hot stellar core, a so-called white dwarf, with about two-thirds the mass of the current Sun, but with a radius only about that of the Earth, i.e., $R \approx R_e \approx 7 \times 10^3$ km $\approx 0.01\, R_\odot$. The surface gravities of white dwarfs are thus typically 10 000 times *higher* than that of the current Sun (Section 19.4).

For "neutron stars," which are the remnants of stars a bit more massive than the Sun, the radius is just about 10 km, more than another factor 500 smaller than white dwarfs (Section 20.3). This implies surface gravities another 5–6 orders of magnitude higher than even white dwarfs. (Imagine what you'd weigh on the surface of a neutron star!)

Since stellar gravities vary over such a large range, it is customary to quote them in terms of the log of the gravity, log g, using CGS units. We thus have gravities ranging from log $g \approx 0$ for red giants, to log $g \approx 4$ for normal stars such as the Sun, to log $g \approx 8$ for white dwarfs, to log $g \approx 13$ for neutron stars. Since the Earth's gravity has a value of log $g_e \approx 3$, the difference of log g from 3 is the number of order of magnitudes more/less that you would weigh on that surface. For example, for neutron stars the difference from Earth is 10, implying you'd weigh 10^{10}, or 10 billion times more on a neutron star! On the other hand, on a red giant, your weight would be about 1000 times *less* than on Earth.

7.2   Surface Escape Speed V_{esc}

Another measure of the strength of a gravitational field is through the surface *escape speed*,

$$V_{esc} = \sqrt{\frac{2GM}{R}} = 618\,\text{km/s}\,\sqrt{\frac{M/M_\odot}{R/R_\odot}}, \tag{7.4}$$

where the latter equality relates this to the escape speed of the Sun. An object of mass m launched with this speed has a kinetic energy $mV_{esc}^2/2 = GMm/R$. This just equals the work needed to lift that object from the surface radius R to escape at a large radius $r \to \infty$,

$$W = \int_R^\infty \frac{GMm}{r^2}\, dr = \frac{GMm}{R}. \tag{7.5}$$

Thus if one could throw a ball (or launch a rocket!) with this speed outward from a body's surface radius R, then by conservation of total energy, that object would reach an arbitrarily large distance from the star, with, however, a vanishingly small final speed.[2]

For the Earth, the escape speed is about 25 000 mph, or 11.2 km/s. By comparison, for the Moon, it is just 2.4 km/s, which is one reason the Apollo astronauts could use a much smaller rocket to get back from the Moon, than they used to get there in the first place. However, escaping from the surface of the Sun (and most any star) is *much* harder, requiring an escape speed of 618 km/s.

7.3   Speed for Circular Orbit

Let us next compare this escape speed with the speed needed for an object to maintain a circular orbit at some radius r from the center of a gravitating body of mass M. For an orbiting body of mass m, we require the gravitational force to be balanced by the centrifugal force from moving along the circle of radius r,

$$\frac{GMm}{r^2} = \frac{mV_{orb}^2}{r}, \tag{7.6}$$

which can be solved to give

$$\boxed{V_{orb}(r) = \sqrt{\frac{GM}{r}}.} \tag{7.7}$$

Note in particular that the orbital speed very near the stellar surface, $r \approx R$, is given by $V_{orb}(R) = V_{esc}/\sqrt{2}$. Thus the speed of satellites in low-Earth orbit (LEO) is about 17 700 mph, or 7.9 km/s.

Of course, orbits can also be maintained at any radius above the surface radius, $r > R$, and Eq. (7.7) shows that, in this case, the speed needed declines as $1/\sqrt{r}$. Thus, for example, the orbital speed of the Earth around the Sun is about 30 km/s, a factor of $\sqrt{R_\odot/\text{au}} = \sqrt{1/215} = 0.0046$ smaller than the orbital speed near the Sun's surface, $V_{orb,\odot} = 434$ km/s.

---

[2] Neglecting forces other than gravity, such as the drag from an atmosphere.

7.4   Virial Theorem for Bound Orbits

If we define the gravitational energy to be zero far from a star, then, for an object of mass m at a radius r from a star of mass M, we can write the gravitational binding energy U at radius r as the *negative* of the escape energy,

$$U(r) = -\frac{GMm}{r}.$$

(7.8)

If this same object is in orbit at this radius r, then the kinetic energy of the orbit is

$$T(r) = \frac{mV_{\text{orb}}^2}{2} = +\frac{GMm}{2r} = -\frac{U(r)}{2},$$

(7.9)

where the second equation uses Eq. (7.7) for the orbital speed $V_{\text{orb}}(r)$. We can then write the *total* energy as

$$E(r) \equiv T(r) + U(r) = -T(r) = \frac{U(r)}{2}.$$

(7.10)

This fact that the total energy E just equals *half* the gravitational binding energy U is an example of what is known as the *Virial Theorem*. It is broadly applicable to most any stably bound gravitational system. For example, if we recognize the thermal energy inside a star as a kind of kinetic energy, it even applies to stars, in which the internal gas pressure balances the star's own self-gravity. This is discussed further in Section 8.2 and the notes on stellar structure in Part II.

7.5   Questions and Exercises

Quick Questions

1. What is the ratio of the *energy* needed to escape the Earth versus that needed to reach LEO? What is the ratio of the associated *speeds*?
2. *Weight change on different bodies*
   Suppose a man weighs 200 lbs. on Earth. What is his weight on:
   a. The Sun?
   b. A red giant with $M = 1M_\odot$ and $R = 100R_\odot$?
   c. A white dwarf with $M = 1M_\odot$ and $R = 0.01R_\odot$?
   d. A neutron star with $M = 1M_\odot$ and $R = 10$ km?
3. In CGS units, the Sun has log $g_\odot \approx 4.44$. Compute the log g for stars with:
   a. $M = 10M_\odot$ and $R = 10R_\odot$.
   b. $M = 1M_\odot$ and $R = 100R_\odot$.
   c. $M = 1M_\odot$ and $R = 0.01R_\odot$.
   d. The Sun has an escape speed of $V_{e\odot} = 618$ km/s. Compute the escape speed V_e of the stars in the above.
4. The Earth orbits the Sun with an average speed of $V_{\text{orb}} = 2\pi$ au/yr $= 30$ km/s. Compute the orbital speed V_{orb} (in km/s) of a body at the following distances from the stars with the quoted masses:

a) $M = 10M_\odot$ and $d = 10$ au.
b) $M = 1M_\odot$ and $d = 100$ au.
c) $M = 1M_\odot$ and $d = 0.01$ au.

5. Rank the following in order of increasing energy change ΔE required:
   a) Escaping the solar system from 1 au.
   b) Escaping the Earth from its surface.
   c) Escaping the Earth from LEO.
   d) Escaping the solar system from Earth–Sun orbit.
   e) Reducing Earth–Sun orbit to an orbit that would impact the Sun.

Exercises

1. *Gravity from Earth versus Moon*
   a. Consider an object on the Earth's surface with the Moon directly above. What is the ratio of the gravity from the Earth versus the Moon, g_e/g_{moon}?
   b. By what factor is the Moon's gravity on the Earth's surface facing the Moon stronger than it is at the Earth's center? Express this symbolically in terms of the Earth's radius R_e, and the Moon's distance d_{em} and mass M_{moon}; then compute its numerical value.
   c. Likewise, compute the analogous ratio between gravity at the Earth's center to a surface point facing directly *away* from the Moon.
   d. Discuss the role the above answers play in the Earth's ocean tides, with particular emphasis on explaining why there are generally two high/low tides per day.

2. *Roche limit*
   Building on the previous exercise, let us next estimate the Roche limit for how close a distance d a moon of mass m and radius r can come to a planet with mass M without breaking up due to tidal forces.
   a. First write down the moon's self-gravity g_m in terms of its mass m and radius r.
   b. Now show that, for the case $d \gg r$, the planet reduces this moon's gravity at a surface point closest to the planet by an amount $g_p \approx 2GMr/d^3$.
   c. Next, set $g_p = g_m$ and solve for d, the Roche distance at which a planet would tidally disrupt a too-close moon.
   d. Given the planet's radius R, recast this alternatively in terms of R and the mean *densities* ρ_m and ρ_p of the moon and planet.
   e. Finally, evaluate the Roche distance d for Earth's Moon, both in km and in Earth radii R_e. (See also Chapter 24, Exercise 2, on the Hill radius for a planet.)

3. *Eclipse*
   a. During a solar eclipse, the Moon just barely covers the visible disk of the Sun. What does this tell you about the relative angular size of the Sun and Moon?
   b. Given that the Moon is at a distance of 0.0024 au, what is the ratio of the *physical* size of the Moon versus the Sun?
   c. Compared with the Earth, the Sun and Moon have gravities of, respectively, $27g_e$ and $g_e/6$. Using this and your answer above, what is the ratio of the *mass* of the Moon to that of the Sun?

    d. Using the above, plus known values for Newton's constant G, Earth's gravity $g_e = 9.8 \text{ m/s}^2$, and the solar radius $R_\odot = 696\,000$ km, compute the masses of the Sun and Moon in kg.

4. *Gravitational escape*
   a) What is the ratio of the *energy* needed to escape the Moon versus the Earth?
   b) What's the ratio for the Sun versus the Earth?
   c) What is the escape speed (in km/s) from a star with:
      i. $M = 10M_\odot$ and $R = 10R_\odot$;
      ii. $M = 1M_\odot$ and $R = 100R_\odot$;
      iii. $M = 1M_\odot$ and $R = 0.01R_\odot$?
   d) To what radius (in km) would you have to shrink the Sun to make its escape speed equal to the speed of light c?
   e) Do you think it possible for a object to have an escape speed of c?

5. *Blowing up a planet*
   In the original *Star Wars* movie, a "Death Ray" is used to blow up the planet Alderaan in just a few (say 10) seconds.
   a) Assuming Alderaan has the same mass M_e and radius R_e as our Earth, estimate its total gravitational binding energy U in joules (J), under the assumption that it has roughly a constant mass density ρ throughout.
   b) Estimate the minimum power P of the Death Ray, giving your answer in both watts (W = J/s), and in solar luminosities, $P/L_\odot$.
   c) Assuming the Death Ray originates from a circular aperature of radius $R = 1$ km, compute its initial energy flux F_a in W/m^2.
   d) What is the ratio of this to the surface flux of the Sun, $F_a/F_\odot$?
   e) What would the temperature T of the Death Ray have to be (in K) if this flux were emitted by blackbody radiation? (Hint: Recall the surface temperature of the Sun is about $T_\odot \approx 6000$ K.)
   f) Assuming the focus of the Death Ray makes this full power just cover uniformly the projected area of Alderaan, compute the associated surface energy flux F_a in W/m^2 on that planet.
   g) What is the ratio of this to the surface flux of the Sun, $F_a/F_\odot$?

8 Stellar Ages and Lifetimes

In our list of basic stellar properties, let us next consider stellar age. Just how old are stars such as the Sun? What provides the energy that keeps them shining? And what will happen to them as they exhaust various available energy sources?

8.1 Shortness of Chemical Burning Timescale for the Sun and Stars

When nineteenth-century scientists pondered the possible energy sources for the Sun, some first considered whether this could come from the kind of chemical reactions (e.g., from fossil fuels such as coal, oil, natural gas, etc.) that power human activities on Earth. But such chemical reactions involve transitions of electrons among various bound states of atoms, and, as discussed in Appendix A (Section A.1) for the Bohr model of the hydrogen atom, the scale of energy release in such transitions is limited to something on the order of 1 eV (electron volt). In contrast, the rest-mass energy of the protons and neutrons that make up the mass is about 1 GeV, or 10^9 times higher. With the associated mass–energy efficiency of $\epsilon \sim 10^{-9}$, we can readily estimate a timescale for maintaining the solar luminosity from chemical reactions,

$$t_{chem} = \epsilon \, \frac{M_\odot c^2}{L_\odot} = \epsilon \, 4.5 \times 10^{20} \, \text{s} = \epsilon \, 1.5 \times 10^{13} \, \text{yr} \approx 15\,000 \, \text{yr}. \tag{8.1}$$

Even in the nineteenth century, it was clear, e.g., from geological processes such as erosion, that the Earth – and so presumably also the Sun – had to be much older than this.

8.2 Kelvin–Helmholtz Timescale for Gravitational Contraction

So let us consider whether, instead of chemical reactions, gravitational contraction might provide the energy source to power the Sun and other stars. As a star undergoes a contraction in radius, its gravitational binding becomes stronger, with a deeper gravitational potential energy, yielding an energy release set by the negative of the change in gravitational potential ($-dU > 0$). If the contraction is gradual enough that the star roughly maintains dynamical equilibrium, then just half of the gravitational energy released goes into heating up the star,[1] leaving the other half available to power the radiative luminosity, $L = -\frac{1}{2}dU/dt$. For a star of observed luminosity L and present-day gravitational binding energy U, we can thus define a characteristic gravitational contraction lifetime,

$$t_{\text{grav}} = -\frac{1}{2}\frac{U}{L} \equiv t_{\text{KH}}, \tag{8.2}$$

where the subscript KH refers to Kelvin and Helmholtz, the names of the two scientists credited with first identifying this as an important timescale. To estimate a value for the gravitational binding energy, let us consider the example for the Sun under the somewhat artificial assumption that it has a uniform, constant density, given by its mass over volume, $\rho = M_\odot/(4\pi R_\odot^3/3)$. Since the gravity at any radius r depends only on the mass $m = \rho 4\pi r^3/3$ *inside* that radius, the total gravitational binding energy of the Sun is given by integrating the associated local gravitational potential $-Gm/r$ over all differential mass shells dm,

$$-U = \int_0^{M_\odot} \frac{Gm}{r} dm = \frac{16\pi^2}{3}G\rho^2 \int_0^{R_\odot} r^4 dr = \frac{3}{5}\frac{GM_\odot^2}{R_\odot}, \tag{8.3}$$

which thus provides an estimate for the Sun's *gravitational binding energy*.

Applying this in Eq. (8.2), we find for the Kelvin–Helmholtz time of the Sun,

$$t_{\text{KH}} \approx \frac{3}{10}\frac{GM_\odot^2}{R_\odot L_\odot} \approx 30\,\text{Myr}. \tag{8.4}$$

Although substantially longer than the chemical burning timescale (8.1), this is still much shorter than the geologically inferred minimum age of the Earth, which is several *billion* years.

8.3   Nuclear Burning Timescale

We now realize, of course, that the ages and lifetimes of stars such as the Sun are set by a much longer *nuclear burning* timescale. As detailed in Chapter 18 (see Figure 18.1), when four hydrogen nuclei are fused into a helium nucleus, the helium mass is about 0.7 percent lower than the original four hydrogen. For nuclear fusion, the above-defined mass–energy burning efficiency is thus now $\epsilon_{\text{nuc}} \approx 0.007$. But in a typical main-sequence star, only some core fraction $f \approx 1/10$ of the stellar mass is hot enough to allow hydrogen fusion. Applying this, we thus find for the nuclear burning timescale

$$t_{\text{nuc}} = \epsilon_{\text{nuc}} f \frac{Mc^2}{L} \approx 10\,\text{Gyr}\,\frac{M/M_\odot}{L/L_\odot}, \tag{8.5}$$

where $\text{Gyr} \equiv 10^9$ yr, i.e., a billion years, or a "giga-year."

---

[1] This is another example of the Virial Theorem for gravitationally bound systems, as discussed in Section 7.4.

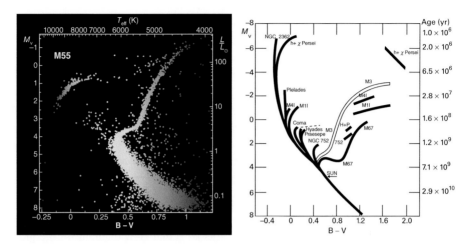

Figure 8.1 Left: H–R diagram for globular cluster M55, showing how stars on the upper main sequence have evolved to lower-temperature giant stars. Credit: NASA/Apod/B.J. Mochejska. Right: Schematic H–R diagram for clusters, showing the systematic peeling off of the main sequence with increasing cluster age. Adapted from A. Sandage, 1957, *Astrophysical Journal* 126, 326.

We thus see that the Sun can live for about 10 Gyr by burning hydrogen into helium in its core. Its present age[2] of ~4.6 Gyr thus puts it roughly halfway through this hydrogen-burning phase, with about 5.4 Gyr to go before it runs out of hydrogen in its core.

8.4 Age of Stellar Clusters from Main-Sequence Turnoff Point

As discussed in Chapter 10 (see Section 10.4 and Eq. (10.11)), observations of stellar binary systems indicate that the luminosities of main-sequence stars scale with a high power of the stellar mass – roughly $L \sim M^3$. In the present context, this implies that high-mass stars should have much shorter lifetimes than low-mass stars.

If we make the reasonable assumption that the same fixed fraction ($f \approx 0.1$) of the total hydrogen mass of any star is available for nuclear burning into helium in its stellar core, then the fuel available scales with the mass, while the burning rate depends on the luminosity. Normalized to the Sun, the main-sequence lifetime thus scales as

$$t_{ms} = t_{ms,\odot} \frac{M/M_\odot}{L/L_\odot} \approx 10 \, \text{Gyr} \left(\frac{M_\odot}{M} \right)^2 . \tag{8.6}$$

The most massive stars, of order $100 \, M_\odot$, and thus with luminosities of order $10^6 L_\odot$, have main-sequence lifetimes of only about about 1 Myr, much shorter the multi-Gyr timescale for solar-mass stars.

---

[2] As inferred, e.g., from radioactive dating of the oldest meteorites.

This strong scaling of lifetime with mass can be vividly illustratied by plotting the H–R diagram of stellar clusters. The H–R diagram plotted in Figure 6.5 is for volume-limited sample near the Sun, consisting of stars of a wide range of ages, distances, and perhaps even chemical composition. But stars often appear in clusters, all roughly at the same distance; moreover, since they likely formed over a relatively short time span out of the same interstellar cloud, they all have roughly the same age and chemical composition. Using Eq. (8.6) together with the the $L \sim M^3$ relation, the age of a stellar cluster can be inferred from its H–R diagram simply by measuring the luminosity L_{to} of stars at the "turnoff" point from the main sequence,

$$ t_{\text{cluster}} \approx 10\,\text{Gyr} \left(\frac{L_\odot}{L_{to}} \right)^{2/3} . \tag{8.7} $$

The left panel of Figure 8.1 plots an actual H–R diagram for the globular cluster M55. Note that stars to the upper left of the main sequence have evolved to a vertical branch of cooler stars extending up to the red giants.[3] This reflects the fact that more-luminous stars exhaust their hydrogen fuel sooner that dimmer stars, as shown by the inverse luminosity scaling of the nuclear burning timescale in Eq. (8.5). The right panel illustrates schematically the H–R diagrams for various types of stellar clusters, showing how the turnoff point from the main sequence is an indicator of the cluster age. Observed cluster H–R diagrams like this thus provide a direct diagnostic of the formation and evolution of stars with various masses and luminosities.

8.5   Questions and Exercises

Quick Questions

1. What are the luminosities (in $L_\odot$) and the expected main-sequence lifetimes (in Myr) of stars with masses:
   a. $10\,M_\odot$?
   b. $0.1\,M_\odot$?
   c. $100\,M_\odot$?
2. Confirm the integration result in Eq. (8.3).
3. What is the age (in Myr) of a cluster with a main-sequence turnoff luminosity at:
   a. $L = 1 L_\odot$?
   b. $L = 10 L_\odot$?
   c. $L = 100 L_\odot$?
   d. $L = 1000 L_\odot$?
   e. $L = 10^5 L_\odot$?
   f. $L = 10^6 L_\odot$?

---

[3] Stars just above this main sequence turnoff are dubbed "blue stragglers." They are stars whose close binary companion became a red giant with a such big radius that mass from its envelope spilled over onto it. This rejuvenated the mass gainer, making it again a hot, luminous blue star.

Exercises

1. Suppose you observe a cluster with a main-sequence turnoff point at a luminosity of $100L_\odot$.

   a. What is the cluster's age, in Myr?

   b. What about for a cluster with a turnoff at a luminosity of $10\,000L_\odot$?

2. *Cluster main-sequence turnoff point*

   A cluster has a main-sequence turnoff at a spectral type G2, corresponding to stars of apparent magnitude $m = +10$.

   a) About what is the luminosity, in $L_\odot$, of the stars at the turnoff point?

   b) About what is the age (in Gyr) of the cluster?

   c) About what is the distance (in pc) of the cluster?

3. *Stellar properties and lifetime*

   A main-sequence star with parallax $p = 0.01$ arcsec has an apparent magnitude $m = +2.5$.

   a. What is its distance d, in pc?

   b. What is its luminosity L, in $L_\odot$?

   c. What is its mass M, in $M_\odot$?

   d. What is its main-sequence lifetime t_{ms}, in Myr?

   e. What is its estimated radius R, in $R_\odot$?

   f. What is its associated surface temperature T, in K?

   g. Suppose this star is in a cluster and sits right below the main-sequence turnoff point in its H–R diagram. What is the age of the cluster, $t_{cluster}$, in Myr?

9 Stellar Space Velocities

The next chapter (Chapter 10) will use the inferred orbits of stars in *binary* star systems to directly determine stellar masses. But first, as a basis for interpreting observations of such systems in terms of the orbital velocity of the component stars, let us review the astrometric and spectrometric techniques used to measure the motion of stars through space.

9.1 Transverse Speed from Proper Motion Observations

In addition to such periodic motion from binary orbits, stars generally also exhibit some systematic motion relative to the Sun, generally with components both transverse (i.e., perpendicular) to and along (parallel to) the observed line of sight. For nearby stars, the perpendicular movement, called "proper motion," can be observed as a drift in the apparent position of the star relative to the more-fixed pattern of more-distant, background stars. Even though the associated physical velocities can be quite large, e.g., $V_t \approx 10–100$ km/s, the distances to stars are so large that proper motions of stars – measured as an angular drift per unit time, and generally denoted with the symbol μ – are generally no bigger than about $\mu \approx 1$ arcsec/yr. But because this is a systematic drift, the longer the star is monitored, the smaller the proper motion that can be detected, down to about $\mu \approx 1$ arcsec/century or less for the most well-observed stars.

Figure 9.1 illustrates the proper motion for Barnard's star, which has the highest μ value of any star in the sky. It is so high, in fact, that its proper motion can even be followed with a backyard telescope, as was done for this figure. This star is actually tracking along the nearly South-to-North path labeled as the "Hipparcos[1] mean" in the figure. The apparent, nearly East–West (EW) wobble is due to the Earth's own motion around the Sun, and indeed provides a measure of the star's parallax, and thus its distance. Referring to the arcsec marker in the lower right, we can estimate the full amplitude of the wobble at a little more than 1 arcsec, meaning the parallax[2] is

---

[1] Hipparcos is an orbiting satellite that, because of the absence of atmospheric blurring, can make very precise "astrometric" measurements of stellar positions, at precisions approaching a milliarcsecond (mas).

[2] Given by half the full amplitude, since parallax assumes a 1 au baseline that is half the full diameter of Earth's orbit.

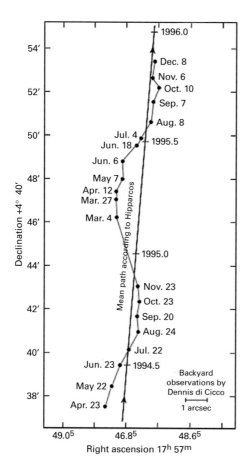

Figure 9.1 Proper motion of Barnard's star. The star is actually tracking along the path labeled as the mean from the Hipparcos astrometric satellite. The apparent wobble is due to the parallax from the Earth's own motion around the Sun. Referring to the lower right label showing 1 arcsec, we can estimate the full amplitude of the parallax wobble as about 1.1 arcsec; but since this reflects a baseline of 2 au from the Earth's orbital diameter, the (1 au) parallax angle is half this, or $p = 0.55$ arcsec, implying a distance of $d = 1/p \approx 1.8$ pc. The figure is taken from the introduction to *The Millennium Star Atlas*, and was produced by Dennis di Cicco for Sky Publishing Corporation.

$p \approx 0.55$ arcsec, implying a distance of $d \approx 1.8$ pc. By comparison, the roughly South-to-North proper motion has a value $\mu \approx 10$ arcsec/yr.

In general, with a known parallax p in arcsec, and known proper motion μ in arcsec/yr, we can derive the associated transverse velocity V_t across our line of sight,

$$V_t = \frac{\mu}{p} \text{ au/yr} = 4.7 \frac{\mu}{p} \text{ km/s}, \tag{9.1}$$

where the latter equality uses the fact that the Earth's orbital speed $V_e = 2\pi$ au/yr $= 30$ km/s. For Barnard's star this works out to give $V_t \approx 90$ km/s, or about three times the Earth's orbital speed around the Sun. This is among the fastest transverse speeds inferred among the nearby stars.

9.2   Radial Velocity from Doppler Shift

We have seen how we can directly measure the transverse motion of relatively nearby, fast-moving stars in terms of their proper motion. But how might we measure the *radial* velocity component *along* our line of sight? The answer is: via the "Doppler effect," wherein such radial motion leads to an observed shift in the wavelength of the light.

To see how this effect comes about, we need only consider some regular signal with period P_o being emitted from an object moving at a speed V_r toward ($V_r < 0$) or away ($V_r > 0$) from us. Let the signal travel at a speed V_s, where $V_s = c$ for a light wave, but might equally as well be the speed of sound if we were to use that as an example. For clarity of language, let us assume the object is moving away, with $V_r > 0$. Then, after any given pulse of the signal is emitted, the object moves a distance $V_r P_o$ before emitting the next pulse. Since the pulse still travels at the same speed, this implies it takes the second pulse an extra time

$$\Delta P = \frac{V_r P_o}{V_s} \tag{9.2}$$

to reach us. Thus the period we observe is longer, $P' = P_o + \Delta P$.

For a wave, the wavelength is given by $\lambda = P V_s$, implying then an associated stretch in the observed wavelength

$$\lambda' = P' V_s = (P_o + \Delta P) V_s = (V_s + V_r) P_o = \lambda_o + V_r P_o, \tag{9.3}$$

where $\lambda_o = P_o V_s$ is the rest wavelength. The associated relative stretch in wavelength is thus just

$$\frac{\Delta \lambda}{\lambda_o} \equiv \frac{\lambda' - \lambda_o}{\lambda_o} = \frac{V_r}{V_s}. \tag{9.4}$$

For sound waves, this formula works in principle as long as $V_r > -V_s$. But if an object moves *toward* us faster than sound ($V_r < -V_s$), then it can basically "overrun" the signal. This leads to strongly compressed sound waves, called "shock waves," which are the basic origin of the sonic boom from a supersonic jet. For some nice animations of this, see:

`www.lon-capa.org/~mmp/applist/doppler/d.htm`

A common example of the Doppler effect in sound is the shift in pitch we hear as the object moves past us. Consider the noise from a car on a highway, for which the "vvvvrrrrrooomm" sound stems from just this shift in pitch from the car engine noise. Figure 9.2 illustrates this for a racing car.

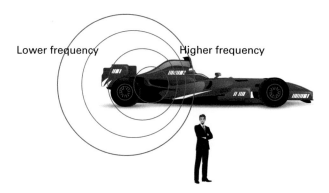

Figure 9.2 Illustration of the Doppler shift of the sound from a racing car.

In the case of light, $V_s = c$, and so we can define the Doppler shift of light as

$$\boxed{\frac{\Delta\lambda}{\lambda_o} = \frac{V_r}{c}; \quad |V_r| \ll c.}$$ (9.5)

This assumes the nonrelativistic case that $|V_r| \ll c$, which applies well to almost all stellar motions. Straightforward observations of the associated wavelengths of spectral lines in the star's spectrum relative to their rest (laboratory measured) wavelengths thus gives a direct measurement of the star's motion toward or away from the observer.

For our above example of Barnard's star, observations of the stellar spectrum show a constant *blueshift* of $\Delta\lambda/\lambda = -3.7 \times 10^{-4}$, implying the star is moving *toward* us, with a speed $V_r = c\Delta\lambda/\lambda = -111$ km/s. This allows us to derive the overall *space velocity*,

$$V = \sqrt{V_r^2 + V_t^2}.$$ (9.6)

For Barnard's star, this gives $V = 143$ km/s, which again is one of the highest space velocities among nearby stars. Mapping the space motion of nearby stars relative to the Sun provides some initial clues about the kinematics of stars in our local region of the Milky Way Galaxy.

9.3   Questions and Exercises

Quick Questions

1. What is the Doppler shift ratio $\Delta\lambda/\lambda_o$ for a star moving away by a speed $V_r = 150$ km/s?
2. A line with rest wavelength $\lambda_o = 1000$ nm is observed in a star to have a wavelength $\lambda = 999$ nm. What is the star's radial speed V_r (in km/s) and is it moving toward or away from us?
3. A star with parallax $p = 0.05$ arcsec has a proper motion $\mu = 1$ arcsec/yr. What is its tangential speed V_t, in both km/s and au/yr?

4. What is the total space velocity V (in km/s) of a star that has both the properties listed in the previous two questions?

Exercises

1. A star with parallax $p = 0.02$ arcsec is observed over 10 years to have shifted by 2 arcsec from its proper motion. Compute the star's tangential space velocity V_t, in km/s.

2. Suppose the star in Exercise 1 has the center of its Balmer-α line at an observed wavelength $\lambda_{obs} = 655.95$ nm.
   a. Is the star moving toward us or away from us?
   b. What is the star's Doppler shift z?
   c. What is the star's radial velocity V_r, in km/s?
   d. What is the star's total *space* velocity V_{tot}, in km/s?

3. An airplane at a distance $d = 10$ km is observed to move across the sky at angular rate of $\mu = 1$ degree/s.
   a. What is the airplane's tangential speed, V_t (in km/hr)?
   b. If a radar gun indicates a radial speed of $V_r = 125$ km/hr, what is the plane's total space velocity, V_{tot} (in km/hr)?
   c. What angle (in degrees) does the plane's velocity make with the observer's line of sight?

4. A police car sitting at a distance b from a highway uses its radar gun to measure the time variation of radial speed $V_r(t)$ of cars passing by, where t is the time relative to the moment when $V_r = 0$.
   a. For cars moving at a constant speed V_c along the highway, derive an expression for $V_r(t)$ in terms of b, V_c and t.
   b. Next, derive a similar formula for the car's angular speed $\mu(t)$ (in degree/s).
   c. Finally, derive an expression for the time variation of the distance $d(t)$ between the police car and the one on the highway.
   d. At what time does $V_r = V_c$?

10 Using Binary Systems to Determine Masses and Radii

Let us next consider how we can infer the masses of stars, namely through the study of stellar *binary systems*.

It turns out, in fact, that stellar binary (and even triple and quadruple) systems are quite common, so much so that astronomers sometimes joke that "three out of every two stars is (in) a binary." The joke here works because often two stars in a binary are so close together on the sky that we can't actually resolve one star from another, and so we sometimes mistake the light source as coming from a single star, when in fact it actually comes from two (or even more). But even in such close binaries, we can often still tell there are two stars by carefully studying the observed spectrum, and, in this case, we call the system a "spectroscopic binary" (see Section 10.2 and Figure 10.2).

But for now, let us first focus on the simpler example of "visual binaries," also known as "astrometric" binaries (see Figure 10.1), since their detection typically requires precise astrometric measurements of small variations of their positions on the sky over time.

10.1 Visual (Astrometric) Binaries

In visual binaries, monitoring of the stellar positions over years and even decades reveals that the two stars are actually moving around each other, much as the Earth moves around the Sun. Figure 10.1 illustrates the principles behind visual binaries. The time it takes the stars to go around a full cycle, called the orbital period, can be measured quite directly. Then if we can convert the apparent angular separation into a physical distance apart – e.g., if we know the distance to the system independently through a measured annual parallax for the stars in the system – then we can use Kepler's third law of orbital motion (as generalized by Newton) to measure the total mass of the two stars.

It is actually quite easy to derive the full formula in the simple case of circular orbits that lie in a plane perpendicular to our line of sight. For stars of mass M_1 and M_2 separated by a physical distance a, Newton's law of gravity gives the attractive force each star exerts on the other,

$$F_\mathrm{g} = \frac{GM_1M_2}{a^2}. \tag{10.1}$$

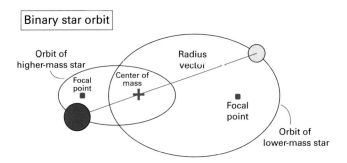

Figure 10.1 Illustration of the properties of a visual binary system.

A key difference from a satellite orbiting the Earth, or a planet orbiting a star, is that, in binary stars, the masses can become comparable. In this case, each star (1, 2) now moves around the *center of mass* at a fixed distance a_1 and a_2, with their ratio given by $a_2/a_1 = M_1/M_2$ and their sum by $a_1 + a_2 = a$. In terms of the full separation, the orbital distance of, say, star 1 is thus given by

$$a_1 = a\frac{M_2}{M_1 + M_2}. \tag{10.2}$$

For the given period P, the associated orbital speeds for star 1 is given by $V_1 = 2\pi a_1/P$. For a stable, circular orbit, the outward centrifugal force on star 1,

$$F_{c1} = \frac{M_1 V_1^2}{a_1} = \frac{4\pi^2 M_1 a_1}{P^2} = \frac{4\pi^2 a}{P^2}\frac{M_1 M_2}{M_1 + M_2}, \tag{10.3}$$

must balance the gravitational force from Eq. (10.1), yielding

$$\frac{G M_1 M_2}{a^2} = \frac{4\pi^2 a}{P^2}\frac{M_1 M_2}{M_1 + M_2}. \tag{10.4}$$

This can be used to obtain the sum of the masses,

$$\boxed{M_1 + M_2 = \frac{4\pi^2}{G}\frac{a^3}{P^2} = \frac{a_{\mathrm{au}}^3}{P_{\mathrm{yr}}^2}M_\odot,} \tag{10.5}$$

where the latter equality shows that evaluating the distance in astronomical units (au) and the period in years gives the mass in units of the solar mass. For a visual binary in which we can actually see both stars, we can separately measure the two orbital distances, yielding the mass ratio $M_2/M_1 = a_1/a_2$. The mass for, e.g., star 1 is thus given by

$$M_1 = \frac{a_{\mathrm{au}}^3}{(1 + a_1/a_2)\,P_{\mathrm{yr}}^2}M_\odot. \tag{10.6}$$

The mass for star 2 can likewise be obtained if we just swap subscripts 1 and 2.

Equations (10.5) and (10.6) are actually forms of Kepler's third law for planetary motion around the Sun. Setting $M_1 = M_\odot$ and the planetary mass M_2, we first note that, for all planets, the mass is much smaller than for the Sun, $M_2/M_1 = a_1/a_2 \ll 1$, implying that the Sun wobbles only slightly (mostly to counter the pull of the most massive planet, namely Jupiter), with the planets thus pretty much all orbiting around the Sun. If we thus ignore M_2 and plug in $M_1 = M_\odot$ in Eq. (10.5), we recover Kepler's third law in (almost) the form in which he expressed it,

$$P_{\mathrm{yr}}^2 = a_{\mathrm{au}}^3. \tag{10.7}$$

To be precise, Kepler showed that, in general, the orbits of the planets are actually *ellipses*, but this same law applies if we replace the circular orbital distance a with the "semimajor axis" of the ellipse. A circle is just a special case of an ellipse, with the semimajor axis equal to the radius.

In general, of course, real binary systems often have elliptical orbits, which, moreover, lie in planes that are not generally normal to the observer line of sight. These systems can still be fully analyzed using the elliptical orbit form of Newton's generalization of Kepler's third law.[1] Indeed, by watching the rate of movement of the stars along the projected orbit, the inclination effect can even be disentangled from the ellipticity.

10.2   Spectroscopic Binaries

As noted, there are many stellar binary systems in which the angular separation between the components is too close to readily resolve visually. However, if the orbital plane is not perpendicular to the line of sight, then the orbital velocities of the stars will give a variable Doppler shift to each star's spectral lines. The effect is greatest when the orbits are relatively *close*, and in a plane *containing* the line of sight, conditions that make such *spectroscopic binaries* complement the wide visual binaries discussed above. Figure 10.2 illustrates the basic features of a spectroscopic binary system.

If the two stars are not too different in luminosity, then observations of the combined stellar spectrum show spectral line signatures of both stellar spectra. As the stars move around each other in such "double-line" spectroscopic binaries,[2] the changing Doppler shift of each of the two spectral line patterns provides information on the changing orbital velocities of the two components, V_1 and V_2.

Considering again the simple case of circular orbits but now in a plane *containing* the line of sight, the inferred radial velocities vary sinusoidally with semi-amplitudes

---

[1] For example, as derived in course notes by Bill Press:
www.lanl.gov/DLDSTP/ay45/ay45c4.pdf

[2] In "single-line" spectroscopic binaries, the brighter "primary" star is so much more luminous that the lines of its companion are not directly detectable; but this secondary star's presence can nonetheless be inferred from the periodic Doppler shifting of the primary star's lines due to its orbital motion.

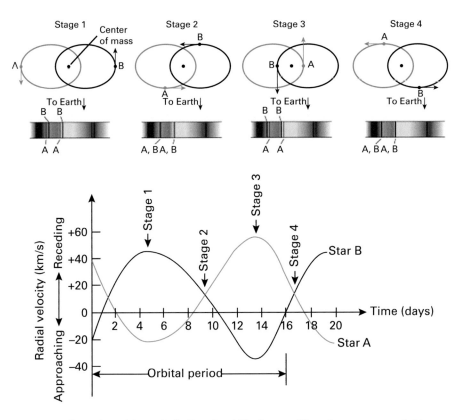

Figure 10.2 Illustration of the periodic Doppler shift of spectral lines in a spectroscopic binary system.

given by the orbital speeds $V_1 = 2\pi a_1/P$ and $V_2 = 2\pi a_2/P$, where a_1 and a_2 are the orbital radii defined earlier. Since the period P is the same for both stars, the ratio of these inferred velocity amplitudes gives the stellar mass ratio, $M_1/M_2 = V_2/V_1$. Using the same analysis as used for visual binaries, but noting now that $a = a_1 + a_2 = PV_1(1 + V_2/V_1)/2\pi$, we obtain a "velocity form" of Kepler's third law given in Eq. (10.6),[3]

$$M_1 = \frac{1}{2\pi G} V_2^3 P (1 + V_1/V_2)^2$$

$$M_1 = \left[\frac{V_2}{V_e}\right]^3 P_{yr} (1 + V_1/V_2)^2 M_\odot, \tag{10.8}$$

---

[3] In the case that the orbital axis in inclined to the line of sight by an angle i, then these scalings generalize with a factor $\sin^3 i$ multiplying the mass, with the velocities representing the inferred Doppler shifted values.

where the latter equality gives the mass in solar units when the period is evaluated in years, and the orbital velocity in units of the Earth's orbital velocity, $V_e = 2\pi$ au/yr $\approx$ 30 km/s. Again, an analogous relation holds for the other mass, M_2, if we swap indices 1 and 2.

10.3   Eclipsing Binaries

In some (relatively rare) cases of close binaries, the two stars actually pass in front of each other, forming an eclipse that temporarily reduces the amount of light we see. Such eclipsing binaries are often also spectroscopic binaries, and the fact that they eclipse tells us that the inclination of the orbital plane to our line of sight must be quite small, implying that the Doppler shift seen in the spectral lines is indeed a direct measure of the stellar orbital speeds, without the need to correct for any projection effect. Moreover, observation of the eclipse intervals provides information that can be used to infer the individual stellar radii.

Consider, for example, the simple case that the orbital plane of the two stars is *exactly* in our line of sight, so that the centers of the two pass directly over each other. As noted, the maximum Doppler shifts of the lines for each star then gives us a direct measure of their respective orbital speeds, V_1 and V_2. In our simple example of circular orbits in Section 10.2, this speed is constant over the orbit, including during the time when the two stars are moving across our line of sight, as they pass into and out of eclipse. In eclipse jargon, the times when the stellar rims just touch are called "contacts," labeled 1–4 for first, second, etc. (see Figure 10.3). Clearly then, once the

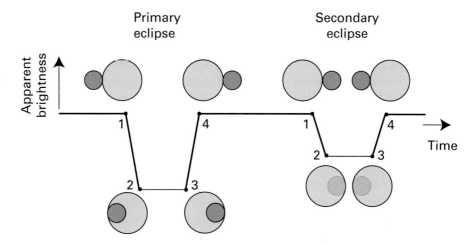

Figure 10.3 Illustration of the how the various contact moments of an eclipsing binary star system correspond to features in the observed light curve.

stellar orbital speeds are known from the Doppler shift, then the radius of the smaller star (R_2) can be determined from the time difference between the first (or last) two contacts,

$$R_2 = (t_2 - t_1)(V_1 + V_2)/2. \qquad (10.9)$$

Likewise, the larger radius (R_1) comes from the time between the second and fourth (or third to first) contacts,

$$R_1 = (t_4 - t_2)(V_1 + V_2)/2. \qquad (10.10)$$

In principle, one can also use the other, weaker eclipse for similar measurements of the stellar radii.

Of course, in general, the orbits are elliptical and/or tilted somewhat to our line of sight, so that the eclipses do not generally cross the stellar centers, but typically move through an off-center chord, sometimes even just grazing the stellar limb. In these cases, information on the radii requires more-complete modeling of the eclipse, and fitting the observations with a theoretical light curve that assumes various parameters. Indeed, to get good results, one often has to relax even the assumption that the stars are spheres with uniform brightness, taking into account the mutual tidal distortion of the stars, and how this affects the brightness distributions across their surfaces. Such details are rather beyond the scope of this general survey course (but could make the basis for an interesting term paper or project).

10.4   Mass–Luminosity Scaling from Astrometric and Eclipsing Binaries

In this simple introduction of the various types of binaries, we have assumed that the orientation, or "inclination" angle i, of the binary orbit relative to our line of sight is optimal for the type of binary being considered, i.e., looking face-on – with $i = 0$ degree inclination between our sight line and the orbital axis – for the case visual binaries; or edge-on – with $i = 90$ degree – for spectroscopic binaries in which we wish to observe the maximum Doppler shift from the orbital velocities. Of course, in practice binaries are generally at some intermediate, often unknown, inclination, leaving an ambiguity in the determination of the mass (typically scaling with $\sin^3 i$) for a given system.

Fortunately, in the relatively few binary systems that are both spectroscopic (with either single or double lines) and either astrometric or eclipsing, it becomes possible to determine the inclination, and so unambiguously to infer the *masses* of the stellar components, as well as the *distance* to the system. Together with the observed apparent magnitudes, this thus also gives the associated luminosities of these stellar components.

Figure 10.4 plots $\log L$ versus $\log M$ (in solar units) for a sample of such astrometric (blue) and eclipsing (red) binaries, showing a clear trend of increasing luminosity with increasing mass. Indeed, a key result is that, for many of the stars (typically those

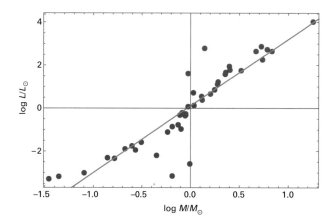

Figure 10.4 A log–log plot of luminosity versus mass (in solar units) for a sample of 26 astrometric (blue, lower points) binaries and 18 double-line eclipsing (red, upper points) binaries. The best-fit line shown follows the empirical scaling, $\log(L/L_\odot) \approx 0.1 + 3.1 \log(M/M_\odot)$.

on the main sequence), the data can be well fit by a straight line in this log–log plot, implying a power-law relation between luminosity and mass,

$$\frac{L}{L_\odot} \approx \left(\frac{M}{M_\odot}\right)^{3.1}. \tag{10.11}$$

In Part II we will use the stellar structure equations for hydrostatic equilibrium and radiative diffusion to explain why the luminosities of main-sequence stars roughly follow this observed scaling with the cube of the stellar mass, $L \sim M^3$ (see Section 17.1).

10.5 Questions and Exercises

Quick Questions

1. An asteroid of mass $m \ll M_\odot$ has a circular orbit at a distance $a = 4$ au from the Sun.
   a. What is the asteroid's orbital period P, in yr?
   b. How would this change for an asteroid with orbital eccentricity $\epsilon = 0.5$ and a semimajor axis $a = 4$ au?
   c. How would this change if asteroid mass is doubled?
   d. What is the period (in yr) for a similar asteroid with $a = 4$ au orbiting a star with mass $M = 4M_\odot$?
2. In eclipsing binaries, note that the net area of stellar surface eclipsed is the same whether the smaller or bigger star is in front. So why then is one of the eclipses deeper than the other? What quantity determines which of the eclipses will be deeper?

3. Approximately what are the luminosities L (in $L_\odot$) of main-sequence stars with the following masses:

   a. $M = 2M_\odot$?
   b. $M = 7M_\odot$?
   c. $M = 10M_\odot$?
   d. $M = 50M_\odot$?
   e. $M = 100M_\odot$?

Exercises

1. *Visual binary*

   Over a period of 10 years, two stars separated by an angle of 1 arcsec are observed to move through a full circle about a point midway between them on the sky. Suppose that over a single year, that midway point is observed itself to wobble by 0.2 arcsec due to the parallax from Earth's own orbit.

   a. How many pc is this star system from Earth?
   b. What is the physical distance between the stars, in au?
   c. What are the masses of each star, M_1 and M_2, in $M_\odot$?

2. *Double-line spectroscopic binary*

   A double-line spectroscopic binary has a spectral line with rest wavelength of 600 nm, which, however, is observed to split into two equal pairs on both the blue and red sides, reaching a maximum wavelength separation of $+/-$ 0.12 nm from the rest wavelength every three months.

   a. What is the orbital speed (in km/s) of the two stars around their common center of mass? (Assume a circular orbit and that we are viewing this system along the stars' orbital plane.)
   b. What is the mass of each star, in units of the Sun's mass $M_\odot$? (Hint: Use the answer to part a and the velocity form of Kepler's third law, in terms of the Earth's orbital speed around the Sun.)
   c. What is the separation between the two stars, in au?

3. *Escape speed and energy*

   a. Recalling that the surface escape speed from the Sun is about 620 km/s, compute V_{esc} from stars with:
      i. $M = 10M_\odot$ and $R = 10R_\odot$
      ii. $M = 1M_\odot$ and $R = 100R_\odot$
      iii. $M = 1M_\odot$ and $R = 0.01R_\odot$.
   b. What is the ratio of the energy E needed to escape the surface of the Moon versus that of the Earth? What's the ratio for the Sun versus the Earth?
   c. Compute the energy E_H for a hydrogen atom with mass m_H to escape from the surface of the Earth. Give your answer in both joules (J) and electron volts (eV).
   d. To what radius R_c (in km) would you have to shrink the Sun to make its escape speed equal to the speed of light c?
   e. What is the significance of this?
   f. Do you think it possible for stars to have $V_{esc} = c$?

4. *Moon's orbit*

   a. Given the Moon's distance from Earth is about $d_m \approx 60\,R_e$, what is the Moon's orbital speed (in km/s)?

   b. What is the ratio of the angular momentum in the Moon's orbit to the angular momentum in the Earth's rotation? (Assume that the moment of inertia of the Earth is given by the formula for a solid, constant-density sphere).

   c. The Moon's tidal effect is slowing the Earth's rotation, transferring its rotational angular momentum to the Moon's orbit. Assuming this continues until they become locked to the same period, use conservation of angular momentum (and Kepler's laws for the Moon's orbit) to derive this period (in days).

   d. What would be the Earth–Moon distance d, relative to its present-day value?

   e. What would be the associated ratio of the angular momentum of the Moon's orbit to the Earth's rotation?

5. *Gravity and circular orbits*

   a. Rederive the expression for the circular orbital velocity as a function of central mass and distance by equating centripetal force to gravitational force for a relatively light body orbiting a much more massive one.

   b. Evaluate the circular orbital velocity V_{leo} of an object in low-Earth orbit (LEO), i.e., orbiting the Earth just above the Earth's atmosphere, say at a height $h \approx 200\,km \ll R_e$ above sea level. Express your answer in km/s and in km/hr.

   c. What is the associated orbital period P_{leo}, expressed in s, hr, and fraction of a day?

   d. What is the ratio of V_{leo} to the rotational velocity of the Earth's equator?

   e. How is this related to P_{leo} evaluated in days? (You can calculate this from your knowledge of the Earth's radius and the length of the day.)

   f. At what radius from the center of the Earth would an object in circular orbit have an orbital period of 1 day, so that it remains fixed at a given position above the Earth's equator? This is the radius R_g for so-called geosynchronous orbit, or "GEO." Express your answer in both km and in Earth radii R_e.

   g. In terms of the Earth's mass M_e and radius R_e, derive an expression for the total energy E_{geo} required to place an object of mass m into GEO.

   h. Evaluate this for a single hydrogen atom of mass m_H, in both joules (J) and electron volts (eV). Compare this with the energy $E_{esc,H}$ required for a hydrogen atom to completely escape the Earth.

Challenge Exercise

6. *Space elevator*

   There are serious engineering proposals to build a "space elevator" centered on a cable that extends from the Earth's equatorial surface radius R_e to the geosynchronous orbit radius R_g computed in the previous exercise.

   a. For a cable with mass-per-length μ, the gravitational weight of the cable section below any given radius r must be supported by the cable's local tension force $T_g(r)$. Ignoring centrifugal forces, but accounting for the radial variation of Earth's gravitational acceleration, derive an expression for $T_g(r)$ for $R_e \leq r \leq R_g$.

b. The strength of cable material can be characterized by its "breaking length" ℓ_b, defined as the length at which an untapered cable will break under its own weight in Earth's constant surface gravity $g_e = 9.8\,\text{m/s}^2$. Compute the breaking length required to support the gravitational tension at GEO, $T_g(R_{geo})$, from above. Write this in terms of Earth's radius R_e.

c. Assuming the cable rotates rigidly at the Earth's rotation period, derive a formula for the cable's rotation speed $V(r)$ in km/s as a function r/R_e. Then use this to derive a similar formula for the associated centrifugal acceleration $V(r)^2/r$, in m/s^2.

d. Now accounting also for this centrifugal acceleration, recompute the net cable tension at GEO $T(R_g)$. Noting that $R_e \ll R_g$, estimate a value for $T(R_g)/T_g(R_g)$. Use this to compute a new value for the required breaking length, ℓ_b, again in R_e.

e. Consider next a strand of cable that consists of a single chain of atoms of mass m_a separated by a distance s, for which $\mu = m_a/s$. For binding energy per atom E_b, the associated force binding them can be approximated by $F \approx E_b/s$ (since energy = force × distance). Derive an expression for the strand's breaking length ℓ_b in terms of E_b, m_a, and g_e.

f. Now use this expression to derive the binding energy per unit mass, E_b/m_a (in eV/m_H), required to support the required tension at GEO $T_g(R_g)$. For $m_a = m_H$, compare this required binding energy E_b to the hydrogen escape energy from the Earth, $E_{esc,H}$, as computed in the previous exercise.

g. Compute the required E_b (in eV) for carbon nanotubes, for which $m_a = m_C = 12m_H$.

11 Stellar Rotation

Let us conclude our discussion of stellar properties by considering ways to infer the rotation of stars. All stars rotate, but in cool, low-mass stars such as the Sun the rotation is quite slow, with, for example, the Sun having a rotation period $P_{rot} \approx$ 26 days, corresponding to an equatorial rotation speed $V_{rot} = 2\pi R_\odot / P_{rot} \approx 2$ km/s. In hotter, more-massive stars, the rotation can be more rapid, typically 100 km/s or more, with some cases (e.g., the Be stars) near the "critical" rotation speed ($\sim$400 km/s) at which material near the equatorial surface would be in a Keplerian orbit! While the rotational evolution of stars is a topic of considerable research interest, it is generally of secondary importance compared with, say, the stellar mass.

11.1 Rotational Broadening of Stellar Spectral Lines

In addition to the Doppler shift associated with the star's overall motion toward or away from us, there can be a *differential* Doppler shift from the parts of the star moving toward and away as the star rotates. This leads to a *rotational broadening* of the spectral lines, with the half-width given by

$$\boxed{\frac{\Delta\lambda_{rot}}{\lambda_o} \equiv \frac{V_{rot}\sin i}{c},}$$
(11.1)

where V_{rot} is the stellar surface rotation speed at the equator, and $\sin i$ corrects for the inclination angle i of the rotation axis to our line of sight. If the star happens to be rotating about an axis pointed toward our line of sight ($i = 0$ degree), then we see no rotational broadening of the lines. Clearly, the greatest broadening is when our line of sight is perpendicular to the star's rotation axis ($i = 90$ degrees), implying $\sin i = 1$, and thus that $V_{rot} = c\Delta\lambda_{rot}/\lambda_o$.

Figure 11.1 illustrates this rotational broadening. The left-side schematic shows how a rotational broadened line profile for flux versus wavelength takes on a *hemispherical* form.[1] For a rigidly rotating star, the line-of-sight component of the surface rotational velocity just scales in proportion to the apparent displacement from the

---

[1] If flux is normalized by the continuum flux F_c, then making the plotted profile actually trace a hemisphere requires the wavelength to be scaled by $\lambda_n \equiv \Delta\lambda_{rot}/r_o$, where $\Delta\lambda_{rot}$ and r_o are the line's rotational half-width and central depth, defined respectively by Eqs. (11.1) and (11.3).

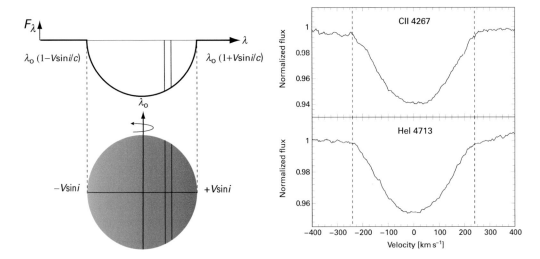

Figure 11.1 Left: Schematic showing how the Doppler shift from rigid-body rotation of a star (bottom) – with constant line-of-sight velocity along strips parallel to the rotation axis – results in a hemispherical line-absorption profile (top). Right: Observed rotational broadening of lines for the star DX Eri, which has an estimated projected rotation speed $V \sin i \approx 240$ km/s. Credit: ESO/T. Rivinius, with permission.

projected stellar rotation axis. Thus for an intrinsically narrow absorption line, the total amount of reduction in the observed flux at a given wavelength is just proportional to the *area* of the vertical strip with a line-of-sight velocity that Doppler-shifts line absorption to that wavelength. As noted above, the total width of the profile is just twice the star's projected equatorial rotation speed, $V \sin i$.

The right panel shows observed rotationally broadened absorption lines for a star with inferred projected equatorial rotation speed $V \sin i \approx 240$ km/s, much higher than ~ 1.8 km/s rotation speed of the solar equator. The flux ratio here is relative to the nearby "continuum" outside the line.

Note that the reduction at line-center is typically only a few percent. This is because such rotational broadening preserves the total amount of reduced flux, meaning that the relative depth of the reduction is diluted when a rapid apparent rotation significantly broadens the line.

A convenient measure for the total line absorption is the"equivalent width,"

$$W_\lambda \equiv \int_0^\infty \left(1 - \frac{F_\lambda}{F_c} \right) d\lambda, \tag{11.2}$$

which represents the width of a "saturated rectangle" with same integrated area of reduced flux (Figure 11.2). For a line with equivalent width W_λ and a rotationally broadened half-width $\Delta\lambda_{\rm rot}$, the central reduction in flux is just

$$r_0 \equiv 1 - \frac{F_{\lambda_0}}{F_c} = \frac{2}{\pi} \frac{W_\lambda}{\Delta\lambda_{\rm rot}}. \tag{11.3}$$

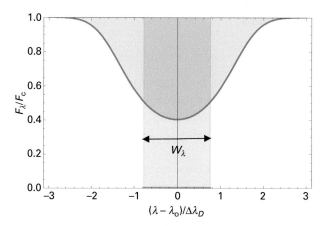

Figure 11.2 Illustration of the definition of the wavelength equivalent width W_λ. The blue curve plots the wavelength variation of the residual flux (relative to the continuum, i.e., F_λ/F_c) for a sample absorption line, with the shaded blue area illustrating the total fractional reduction of continuum light. The tan plot show a box profile with width W_λ, defined such that the total tan shaded area is the same as the blue area for the line profile.

For example, for the HeI 471.3 nm line plotted in the lower box in the right panel of Figure 11.1, the central reduction is just $r_0 \approx 1 - 0.96 = 0.04$, while the velocity half-width (given, e.g., by the vertical dotted lines) is $V \sin i \approx 275$ km/s, corresponding to a wavelength half-width $\Delta\lambda_{\rm rot} \approx 0.43$ nm. This implies an equivalent width $W_\lambda \approx 0.027$ nm, or about 17 km/s in velocity units.

11.2 Rotational Period from Starspot Modulation of Brightness

When Galileo first used a telescope to magnify the apparent disk of the Sun, he found it was not the "perfect orb" idealized from antiquity, but instead had groups of relatively dark "sunspots" spread around the disk. By watching the night-to-night migration of these spots from the East to the West, he could see directly that the Sun is rotating, with a mean period of about 25 days.[2]

Although other stars are too far away to resolve directly the stellar disk and thus make similar direct detections of analogous "starspots," in some cases such spots are large and isolated enough that careful photometric measurement of the apparent stellar brightness shows a regular modulation over the stellar rotation period P.

If the star also shows rotationally broadened spectral lines with an associated inferred projected rotational speed $V_{\rm rot} \sin i$, then the basic relation $V_{\rm rot} = 2\pi R/P$ implies a constraint on the minimum possible value for the stellar radius, $R_{\rm min} = V_{\rm rot} \sin i \, P/2\pi$.

---

[2] Actually, the Sun does not rotate as a rigid body, but has about 10 percent faster rotation at its equator than at higher latitudes; see Chapter 14 and Figure 14.6.

11.3   Questions and Exercises

Quick Questions

1. A line with rest wavelength $\lambda_o = 500$ nm is rotational broadened to a full width of 0.5 nm. Compute the value of $V \sin i$, in km/s.
2. Derive Eq. (11.3) from the definitions of rotational Doppler width $\Delta\lambda_{rot}$ (Eq. (11.1)) and equivalent width W_λ (Eq. (11.2)), using the wavelength scaling given in footnote 1.
3. Write a formula for the equatorial rotation speed V_{rot} (in km/s) in terms of a star's rotation period (in days) and radius R (in $R_\odot$).
4. If the equatorial rotation speed is V_{rot}, what is the associated speed at any latitude ℓ? Does it make any difference if this is in the northern ($\ell > 0$) or southern ($\ell < 0$) hemisphere?

Exercises

1. *Rotational line broadening*
   The Balmer-α spectral line in a star is found to have a rotationally broadened width of $\Delta\lambda_{rot} = 0.2$ nm.
   a. What is the projected rotation speed $V_{rot} \sin i$, in km/s?
   b. If the line has an equivalent width $W_\lambda = 0.01$ nm, what is the line's central depth factor r_o?
   c. If this star also shows periodic variation from starspots with a period $P = 2$ days, what is the minimum stellar radius, R_{min}, in both km and $R_\odot$?
2. *Spot variation of brightness*
   A star with radius $R = 2R_\odot$ is observed to show a magnitude variation $\Delta m = 0.1$ with a period $P = 2$ days.
   a. What is the associated fractional variation in the observed stellar flux, $\Delta f/f$?
   b. If this variation is caused by single circular starspot that has a temperature half that of the rest of the star, what is that spot's fractional area, $A_{spot}/\pi R^2$?
   c. What is the spot's diameter, D_{spot}, in km?
   d. What is the star's equatorial rotation speed V_{rot}, in km/s?
   e. If the star is viewed from an inclination $i = 30$ degree, what would be the rotational broadening, $\Delta\lambda_{rot}$, of its Balmer-α line, in nm?
3. *Critical rotation*
   A star with equatorial radius R and mass M rotates with a period P.
   a. What is the star's equatorial rotation speed, V_{rot}?
   b. What is the star's equatorial orbital speed, V_{orb}?
   c. Defining the ratio $W \equiv V_{rot}/V_{orb}$, derive an expression for the critical period P_{crit} at which $W = 1$.
   d. What do think will happen if a star approaches $W \to 1$?
   e. Be stars are main-sequence B stars that show emission lines arising from a circumstellar disk of gas. Discuss how this might be related to the fact that many of these stars are inferred to have $W > 0.9$.

4. *Launch speed into Earth orbit*
   a. Referring to the previous exercise, what is $W \equiv V_{rot}/V_{orb}$ for the Earth?
   b. In terms of V_{orb} and W, compute the change in speed ΔV needed to launch a satellite from the equator into Earth orbit in the direction (eastward) of its rotation.
   c. How does this differ from the speed to launch in the polar direction (north/south)?
   d. Derive an expression for the variation of an eastward launch speed ΔV as a function of latitude ℓ.

12 Light Intensity and Absorption

12.1 Intensity versus Flux

Our initial introduction of surface brightness characterized it as a flux confined within an observed solid angle, F/Ω. But actually the surface brightness is directly related to a more general and fundamental quantity known as the *specific*[1] *intensity, I*. In the exterior of stars, the intensity is set by the surface brightness $I \approx F/\Omega$, but it can also be specified in the stellar *interior*, where it characterizes the properties of the radiation field as energy generated in the core is transported to the surface.

A simple analog on Earth would be an airplane flying through a cloud. Viewed from outside, the cloud has a surface brightness from reflected sunlight, but as the plane flies into the cloud, the light becomes a "fog" coming from all directions, with the specific intensity in any given direction depending on the details of the scattering through the cloud.

Formally, intensity is defined as the radiative energy per unit area and time that is pointed into a specific *patch of solid angle $d\Omega$* centered on a specified direction. The left side of Figure 12.1 illustrates the basic geometry. As the solid angle of the projected emitting area declines with the inverse square of the distance, $\Omega_{\text{em}} = A_{\text{em}} \cos \theta / d^2$, the fixed solid angle receiving the intensity grows in area in proportion to the distance-squared, $A_{\text{rec}} = \Omega_{\text{rec}} d^2$. In essence, the two distances cancel, and so the *intensity remains constant with distance*.

In this context it is perhaps useful to think of intensity in terms of a narrow beam of light in a particular direction – like a laser "beam" – whereas the flux depends on just the total amount of light energy that falls on a given area of a detector, regardless of the original direction of all the individual "beams" that this might be made up of. However, while valid, this perspective might suggest that intensity is a vector and flux is a scalar, whereas in fact the *opposite* is true. The intensity has a directional dependence through the specification of the direction of the solid angle being emitted into, but it itself is a scalar! The flux measures the rate of energy (a scalar) through a given area, but this has an associated direction given by the normal to that surface area; thus the flux is a vector, with its three components given by the three possible orientations of the normal to the detection area.

---

[1] Often this "specific" qualifer is dropped, leaving just "intensity."

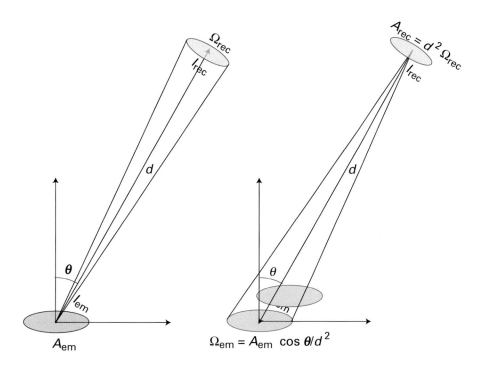

Figure 12.1 Left: The intensity I_{em} emitted into a solid angle Ω_{rec} located along a direction that makes an angle θ with the normal of the emission area A_{em}. Right: The intensity I_{rec} received into an area $A_{rec} = d^2\Omega_{rec}$ at a distance d from the source with projected solid angle $\Omega_{em} = A_{em}\cos\theta/d^2$. Since the emitted and received energies are equal, we see that $I_{em} = I_{rec}$, showing that intensity is invariant with distance d.

For known specific intensity $I(\hat{n})$ in any direction $\hat{n}$, the vector flux $\vec{F}$ is given by vector integration over solid angle Ω,

$$\vec{F} = \oint I(\hat{n})\,\hat{n}\,d\Omega. \tag{12.1}$$

For stars in which the emitted radiation is, at least to a first approximation, spherically symmetric, the only nonzero component of the flux is along the radial direction away from the star. If the angle between any given intensity beam I with the radial direction is written as θ, then its contribution to the radial flux is proportional to $I\cos\theta$; the total radial flux is then obtained by integrating this contribution over solid angle,

$$F = \int I(\theta)\cos\theta\,d\Omega = 2\pi \int_0^{\pi} I(\theta)\cos\theta\sin\theta d\theta. \tag{12.2}$$

As discussed in Section 2.4, and illustrated in Figure 2.5, we can write the differential solid angle as $d\Omega = \sin\theta d\theta d\phi$, with ϕ the azimuthal angle (analogous to longitude on the Earth). The second equality integrates over this azimuth to give the factor 2π.

As a simple example, let us assume the Sun has a surface brightness $I_{\odot}$ that is constant, both over its spherical surface of radius $R_{\odot}$, and also for all *outward* directions from the surface.[2] Now consider the flux $F(d)$ at some distance d (for example at Earth, for which $d = 1$ au). At this distance, the visible solar disk has been reduced to an angular radius $\theta_d = \arcsin(R_\odot/d)$, so that the angle range for the nonzero local intensity has shrunk to the range $0 < \theta < \theta_d$, i.e.,

$$I(\theta) = I_\odot; \quad 0 < \theta < \theta_d$$
$$= 0; \quad \theta_d < \theta < \pi. \tag{12.3}$$

Noting that $\cos \theta_d = \sqrt{1 - R_\odot^2/d^2}$, we then see that evaluation of the integral in Eq. (12.2) gives for the flux

$$F(d) = \pi I_\odot (1 - \cos^2 \theta_d) = \pi I_\odot \frac{R_\odot^2}{d^2}. \tag{12.4}$$

Again, within the cone of half-angle θ_d around the direction toward the Sun's center, the observed intensity is the same as at the solar surface $I = I_\odot$. But the shrinking of this cone angle with distance gives the flux an inverse-square dependence with distance, $F(d) \sim 1/d^2$.

To obtain the flux at the surface radius R of a blackbody, we note that $I = B(T)$ for outward directions with $0 < \theta < \pi$, but is zero for inward directions with $\pi/2 < \theta < \pi$. Noting then that $\sin \theta \, d\theta = -d \cos \theta$, we can readily carry out the integral in Eq. (12.2), yielding the Stefan–Boltzmann law (cf. Eq. 5.2) for the radially outward surface flux

$$F_* \equiv F(R) = \pi B(T) = \sigma_{sb} T^4. \tag{12.5}$$

This also follows from the general flux scaling given in Eq. (12.4) if we just set $d = R_\odot$ and $I_\odot = B(T)$.

12.2   Absorption Mean-Free-Path and Optical Depth

The light we see from a star is the result of competition between thermal emission and absorption by material within the star. Let us first focus on the basic scalings for the absorption by considering the simple case of a beam of intensity I_o along a direction z perpendicular to a planar layer that consists of a local number density $n(z)$ of absorbing particles of projected cross-sectional area σ (see Figure 12.2.) We can characterize the mean-free-path that light can travel before being absorbed within the layer as

$$\ell \equiv \frac{1}{n \sigma} = \frac{1}{\rho \kappa}. \tag{12.6}$$

The latter equality instead uses the mass density $\rho = \mu n$, where μ is the mean mass of stellar material per absorbing particle. The cross section divided by this mass defines

---

[2] Actually, the light from the Sun is "limb darkened," meaning the intensity directly upward is greater than that at more oblique angles toward to the local horizon, or limb. See Section D.2.

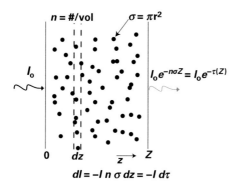

Figure 12.2 Illustration of the attenuation of an intensity beam I_0 by a planar layer of absorbing particles with cross section σ and number density n.

what is called the *opacity*, $\kappa \equiv \sigma/\mu$, which is thus simply the cross section per unit mass of the absorbing medium.

Within a narrow (differential) layer between z and $z + dz$, the probability of light being absorbed is just $d\tau \equiv dz/\ell$. This implies an associated fractional reduction $dI/I = -d\tau$ in the local intensity $I(z)$. We can thus write this change in intensity in terms of a simple differential equation,

$$\frac{dI}{dz} = -\kappa\rho I \quad \text{or} \quad \frac{dI}{d\tau} = -I. \tag{12.7}$$

Straightforward integration using the boundary condition $I(z = 0) = I_0$ at the layer's leading edge at $z = 0$ gives

$$\boxed{I(z) = I_0 e^{-\tau(z)},} \tag{12.8}$$

where

$$\tau(z) \equiv \int_0^z \frac{dz'}{\ell} = \int_0^z n(z')\sigma\,dz' = \int_0^z \kappa\rho(z')\,dz' \tag{12.9}$$

represents the integrated *optical depth* from the surface to a position z within the layer. It is clear from its initial definition that one can think of optical depth as simply the number of mean-free-paths between two locations.

12.3 Interstellar Extinction and Reddening

One practical example of such exponential reduction of light by absorption is the case of interstellar "extinction" of starlight. As detailed in Chapter 21, the space between stars – called the *interstellar medium* (ISM) – is not completely empty, but contains a certain amount of gas and dust. Compared to a stellar atmosphere, or indeed even to a strong terrestrial vacuum, the density is very small, often only a few atoms per cubic centimeter, or a few hundred dust particles per cubic kilometer. But over the huge distances between stars, the associated optical depth τ for extinction of the star's light by scattering and/or absorption can become quite significant, leading to a substantial reduction in the star's apparent brightness.

For a star of radius R and surface intensity I, the luminosity is $L = 4\pi^2 R^2 I$, and in the absence of any absorption the *intrinsic* flux at a distance d is just $F_{int}(d) = L/4\pi d^2 = \pi I (R/d)^2$. But in the case with ISM absorption, the *observed* flux is again (cf. Eq. (12.8)) reduced by the optical depth exponential absorption factor

$$F_{obs}(d) = F_{int}(d)e^{-\tau}, \tag{12.10}$$

where the subscripts stand for "observed" and "intrinsic." The level of this ISM absorption can also be characterized in terms of the number of *magnitudes of extinction*,

$$A \equiv m_{obs} - m_{int} = 2.5 \log \left(\frac{F_{int}}{F_{obs}} \right) = 2.5 \, \tau \log e \approx 1.08 \, \tau. \tag{12.11}$$

In interpreting the observed magnitude of a "standard candle" star with known luminosity, the failure to account for any such extinction can lead to an inferred distance d_{inf} that overestimates the star's true distance d. For observations in the visual band V, we can define an associated visual extinction $A_V \equiv V_{obs} - V_{int} \approx 1.08\tau_V$, where τ_V is the optical depth within the visual band.

In practice, interstellar extinction is generally dominated by the opacity associated with interstellar grains of dust. For large dust grains, the absorption cross section just depends on the physical size, for example given by $\sigma = \pi r^2$ for spherical grains of radius r.

But interstellar dust grains are often very tiny, even microscopic, with sizes of less than 1 μm (micron), and so comparable to the wavelength of optical light. For light in the red or infrared that has a wavelength larger than the dust size, $\lambda > r$, the effective cross section, and thus the associated dust opacity, is reduced, because, in a loose sense, the dust particle can only interact with a fraction of the light wave. Because this redder, longer-wavelength light is less strongly absorbed than the bluer, shorter wavelengths, the remaining light tends to appear "reddened," much in the same way as the Sun's light at sunset.

This reddening can be quantified in terms of a formal *color excess*, defined in terms of the standard B and V filters of the Johnson photometric system,

$$E_{B-V} \equiv (B-V)_{obs} - (B-V)_{int}. \tag{12.12}$$

This color excess tends to increase with increasing visual extinction magnitude $A_V \equiv V_{obs} - V_{int}$. If the intrinsic colors are known (e.g., from the star's spectral type), then, for a given model of the wavelength dependence of the opacity, measuring this color excess makes it possible to estimate of the visual extinction magnitude A_V. Among other things, this allows one to reduce or remove the error in determining the stellar distance.

The detailed variation of dust opacity depends on the size, shape, and composition of the dust, but often it is approximated as scaling as an inverse power law in wavelength, i.e.,

$$\kappa(\lambda) \sim \lambda^{-\beta},$$

where the power index (also known as the "reddening exponent") ranges from $\beta \approx 1$ for "Mie scattering" to $\beta \approx 4$ for "Rayleigh scattering."

The latter is a good approximation for scattering by air molecules and dust in the Earth's atmosphere. The scattering of blue light out of the direction from the Sun makes the sunset red, while all that scattered blue light makes the sky blue.

For ISM dust, the weaker $\beta \approx 1$ scaling is more appropriate, but even this can make a marked difference in the level of extinction for different wavelengths. Details are discussed in Section 21.3 on dust extinction by Giant Molecular Clouds of the ISM.

12.4 Questions and Exercises

Quick Questions

1. *"Opacity" of people versus electrons*
   a. Seen standing up, about what is the cross section (in cm^2) of a person with height 1.8 m and width 0.5 m?
   b. If this person has a mass of 60 kg, what is his/her "opacity" $\kappa = \sigma/m$, in cm^2/g?
   c. How does this compare with the (Thomson) electron scattering opacity κ_e for a fully ionized gas with solar abundances?
2. Derive expressions for d_{inf}/d in terms of both the absorption magnitude A and the optical depth τ.
3. What is the electron scattering opacity κ_e for a fully ionized medium with the mass fractions of hydrogen, helium, and metals given by $X = 0$, $Y = 0.98$, and $Z = 0.02$?
4. What is the extinction magnitude A for a star behind an interstellar cloud with optical thickness $\tau = 4$?
5. Suppose dust absorption has a reddening exponent $\beta = 1$. Write a formula for the ratio of extinction magnitude A_2/A_1 between that at a wavelength λ_1 and that at $\lambda_2 = f\lambda_1$.

Exercises

1. An interstellar cloud of size $D = 10\,pc$ contains dust that gives light a mean-free-path $\ell = 5\,pc$.
   a. What is the cloud's optical depth, τ?
   b. What is the associated absorption magnitudes of extinction, A?
   c. For a star behind the cloud, what is the ratio of observed to intrinsic brightness, F_{obs}/F_{int}?

d. If one did not know about this absorption, by what factor d_{inf}/d would one overestimate the star's distance?

2. *Exponential atmosphere*

As detailed in Section 15.1, often in an atmosphere of a planet or star, the mass density ρ declines exponentially with height z,

$$\rho(z) = \rho_0 \exp^{-z/H}, \qquad (12.13)$$

where H is called the (density) scale height, and $\rho_0 \equiv \rho(z = 0)$ is the density at some reference level (e.g., sea level on Earth, or the "photosphere" for the Sun) where $z \equiv 0$.

a. Assuming a fixed opacity κ, compute the optical depth $\tau(z)$ from any height z to an external observer looking vertically down into the atmosphere from large distances $z \to \infty$ outside it.

b. If someone at height z shines a flashlight with intensity I_0 vertically through this atmosphere, what is the intensity I_∞ such an external observer would see?

c. Now consider an external star with flux F_* that shines from directly above (called the local *zenith*) down to an observer at height z. What flux $F(z)$ does that observer see?

d. Now suppose the star's direction makes an angle θ with the zenith; derive the observer flux $F(z, \mu)$, where $\mu \equiv \cos\theta$.

e. Finally, derive an associated expression for the star's change in magnitude $\Delta m(z, \mu)$ from this atmospheric absorption.

3. *Dust-grain properties*

a. Suppose spherical dust grains have a radius $r = 0.1\,\text{cm}$ and individual mass density $\rho_g = 1\,\text{g/cm}^3$. What is their cross section σ, mass m, and associated opacity κ?

b. If the number density of these grains is $n_d = 1\,\text{cm}^{-3}$, what is the mass density of dust ρ_d and the mean free path ℓ for light?

c. What is the optical depth at a physical depth 1 m into a planar layer of such dust absorbers?

d. What fraction of impingent intensity I_0 makes it to this depth?

4. *Absorption by coal dust.*

Imagine you are a coal miner working under a bright, 1400 W lamp that is at a distance of 10 m away from your work location.

a. Assuming the lamp emits its light isotropically, what is the flux of light on your workspace, in W/m²? (Assume the walls of the coal mine are perfect absorbers, i.e., with zero albedo). How does this compare with the flux of sunlight on a sunny day at the surface?

b. Now suppose there is a cave-in that fills the mine with black, spherical coal dust particles of diameter 0.1 mm, and with a uniform number density of 20 particles per cubic centimeter. What is the mean-free-path ℓ (in m) of light in the mine?

c. What is the optical depth τ between you and the lamp?

d. Now what is the flux on your workspace?

e. What distance had this flux value before the cave-in?

    f. How close would you have to move the lamp to make the flux on your workspace be the same as before the cave-in?

5. *Absorption by interstellar dust*

    Imagine you are an astronomer observing a star with luminosity $1000L_\odot$ that is at a distance of 10 pc.

    a. What is the flux of light you observe? Give your answer in both W/m^2 and $L_\odot/pc^2$.

    b. Now assume the space between you and the star contains spherical dust particles of diameter 1 μm and number density of 6000 particles per cubic kilometer. Assuming the dust absorbs or scatters light in proportion to its geometric cross section, what is the optical depth between you and the star?

    c. What is now the flux you observe? Again give your answer in both W/m^2 and $L_\odot/pc^2$.

    d. If you knew the star's luminosity but didn't know about the dust, what distance (in pc) would you infer for the star based on your observed flux?

    e. What is the change in the apparent magnitude of this star resulting from the dust absorption?

6. *Uniform, planar, radiating slab*

    Consider a uniform planar slab that is infinite in both horizontal directions (x, y) but has a total thickness Z in the vertical direction z. Suppose the uniform mass density ρ comes from purely absorbing particles of mass m and radius r. Write expressions for the:

    a. particle number density n;

    b. particle absorption cross section σ;

    c. opacity of the medium κ;

    d. photon mean-free-path ℓ;

    e. total slab (vertical) optical thickness τ.

    f. Next, suppose a heating element located along the slab midplane at $z = Z/2$ generates an energy flux F (energy/area/time). Assuming the slab is very optically thick ($\tau \gg 1$), derive expressions for:

       i. The total path length L that photons generated at the center ($z = Z/2$) must travel before escaping out the bottom or top (at $z = 0, Z$).

       ii. $T_s = T(z = 0) = T_s(z = Z)$, the temperature at the lower and upper surfaces at $z = 0, Z$.

       iii. $T_c = T(z = Z/2)$, the temperature at the slab center.

13 Observational Methods

13.1 Telescopes as Light Buckets

Stars are so far away that, of the several hundred *billion* in our Galaxy, only about 5000 are visible to our naked eye. Even when the pupils in our eye are dark adapted, they have a maximum diameter of only about 7 mm, limiting the light reaching our retina. Telescopes provide a way to greatly improve on this by collecting the light from a much greater aperture, effectively acting as "light buckets." For a circular aperture of diameter D, the amount of light gathered scales in proportion to the collection area,

$$A = \frac{\pi}{4}D^2. \tag{13.1}$$

As illustrated in Figure 13.1, telescopes can generally be categorized as *refractors* versus *reflectors*. Much like our eyes, refractor telescopes use a lens to bend incoming light into a focus; but such lenses can have a diameter up to about 100 times larger than our pupils, thus collecting 10 000 times as much light. For even larger lenses, housing the long focal length becomes unwieldy, and so this is near the practical upper limit for refractor telescopes.

But reflector telescopes, wherein light is collected by a large primary mirror, can be built with much larger total apertures. The largest optical reflectors currently in operation have diameters about 10 m,[1] formed by combining approximately 100 meter-size hexagonal mirror segments. For example, a 7 m diameter mirror that is now a thousand times the aperture of our pupil can collect a *million* times as much light as our eyes.

Using the definition of magnitude from Section 3.3, this can be used to derive a general formula for the increase in limiting magnitude resulting just from an increased aperture,

$$m_{\text{lim}} \approx 7.5 + 5\log(D/\text{cm}). \tag{13.2}$$

Equation (13.2) does apply directly to amateur telescopes that are used to view the sky by eye. But, in practice, large research telescopes use modern digital camera detectors with efficiencies that well exceed that of our retina. By also integrating the exposure for many minutes or hours, they can detect much fainter objects with magnitudes much larger than the aperture-based limit (Eq. (13.2)). (See Exercise 2 at the end of

---

[1] The Extremely Large Telescope (ELT) currently under construction in the Atacama desert in Chile will have diameter of 39 m! First light is planned for 2025.

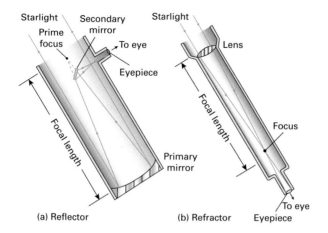

Figure 13.1 Illustration of the basic differences between (a) reflecting and (b) refracting telescopes. Credit: Pearson Prentice Hall.

this chapter.) In practice, the limit is often set by the background darkness of the local sky, one reason modern telescopes are built at remote sites, well away from the light pollution of cities, or placed into orbit, like the Hubble Space Telescope (HST), for which the lack of atmosphere allows for an extremely dark background. The HST primary mirror has only a modest diameter of 2.4 m, but its placement in orbit, above the atmosphere, makes it capable of deep exposure observations that larger, ground-based telescopes cannot match. See, for example, the illustration of the Hubble Ultra Deep Field in Figure 29.5 and the associated discussion in Section 29.4.

13.2 Angular Resolution

Another advantage to a large mirror diameter is that it enables a higher angular resolution. For light of wavelength λ, the diffraction from a telescope with diameter D sets a fundamental limit to the smallest possible angular separation that can be resolved,

$$\alpha = 1.22 \frac{\lambda}{D} = 0.25 \, \text{arcsec} \, \frac{\lambda/\mu\text{m}}{D/\text{m}} = 2.5 \, \text{arcsec} \, \frac{\lambda/\text{cm}}{D/\text{km}}, \qquad (13.3)$$

where the latter two equalities are scaled respectively for optical and radio telescopes.

For ground-based optical telescopes, this ideal diffraction limit is not generally reached, because turbulence in the Earth's atmosphere blurs the image over ~ 1 arcsec or more, an effect known as "astronomical seeing." But this can be reduced to resolutions approaching 0.1 arcsec through a technique called *adaptive optics*, wherein reflection from a laser beam shot up into the sky is used to estimate these seeing distortions, and then dynamically deform secondary mirrors to correct for them.

The laser from the the Very Large Telescope (VLT) array in Figure 26.1 is an example of such adaptive optics in action.

The sharpest focus requires the primary mirror to have a *parabolic* shape. The primary mirror of the Hubble Space Telescope (HST) was mistakenly (and quite infamously) ground instead to a spherical form, leading to a "spherical aberration" in images that had to be subsequently corrected by secondary optics. But with this correction, and despite the modest 2.4 m diameter of its primary mirrors, HST's location in orbit above atmospheric distortions and light pollution has helped it to revolutionize observational astronomy.[2]

13.3 Radio Telescopes

Since radio waves can propagate through the clouds that block visible light, large radio telescopes have been constructed even in locations with poor weather conditions. The radio reflector is now called a "dish," with the largest ones (e.g., the 300 m Arecibo telescope in Puerto Rico) built into natural depressions in the terrain, extending over hundreds of meters. Such dishes are not steerable, but by positioning the receiver around the focal plane they can effectively aim at a range of positions within 30 degrees from the local zenith. The largest steerable dishes range up to 100 m in diameter.

The Very Large Array (VLA) in New Mexico consists of 27 individual dishes that are each 25 m in diameter, positioned on tracks that can spread them over a baseline of up to 30 km. While the sensitivity is set by the combined collective area of the many dishes, a technique called *interferometry* combines their signals to give angular resolution associated with this wider baseline. It observes at wavelengths from 0.6 cm to 410 cm.

An extension of this technique, called Very Long Baseline Interferometry (VLBI) can even combine signals from telescopes spread all around the globe; their diffraction limit can thus, in principle, approach that of a telescope the size of the entire Earth! An impressive recent example is the Event Horizon Telescope (EHT), which used an array of two dozen telescopes to image the millimeter-wavelength emission around a black hole, with angular resolution near 25 *micro*-arcsec! See Figure 26.8 and Exercise 4 in Section 26.5.

The Atacama Large Millimeter Array (ALMA) consists of 66 antennas spread over up to 16 km of the very dry Atacama desert in Chile; the limited water vapor reduces the absorption of millimeter and submillimeter waves enough to allow detection in this intermediate waveband, which is key for, e.g., diagnosing conditions in star-forming regions that have many magnitudes of extinction (Sections 12.3 and 21.3, and Chapter 22) at shorter wavelengths in the visible.

---

[2] Particularly noteworthy are the weeklong exposures allowed by its uniquely dark sky background; known as the Hubble Deep Fields, these exposures revealed huge numbers of very faint, very distant galaxies up to 10 Gly away. See Section 29.4 and Figure 29.5.

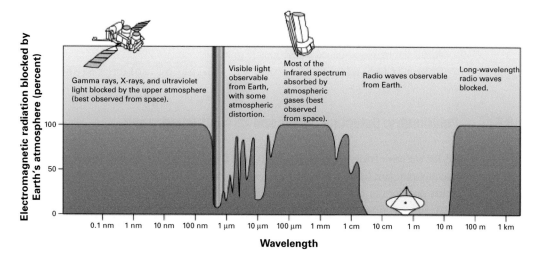

Figure 13.2 The percentage of electromagnetic radiation that is blocked by Earth's atmosphere, plotted as function of wavelength. Image credit: NASA.

13.4 Space-Based Missions

More generally, as illustrated in Figure 13.2, Earth's atmosphere effectively *blocks* radiation in some spectral bands, e.g., at shorter wavelengths ($\lambda < 350$ nm) below the visible. Thus observations in these ultraviolet (UV), X-ray, and gamma-ray wavebands can only be made from orbiting spaced-based platforms above the atmosphere.

The HST has been a principal instrument in the near UV (100 nm $< \lambda < 400$ nm), allowing improved study of hot stars and warm interstellar gas with temperatures $10\,000$ K $< T < 100\,000$ K. In the far and extreme UV (10 nm $< \lambda < 92.1$ nm), ionization by hydrogen in the local interstellar medium largely attenuates radiation from any more distant sources.

At X-ray wavelengths (0.10 nm $< \lambda < 10$ nm), corresponding to high-energy photons (0.1 keV $< E < 100$ keV), telescopes probe very energetic regions with temperatures heated to millions of kelvin, e.g., from accretion onto compact objects such as neutron stars and black holes (Section 20.5), or hot interstellar bubbles that are shock heated by supernova explosions (Section 21.2).

At still shorter, gamma-ray wavelengths $\lambda < 0.01$ nm, with still higher photon energies ($E >$ MeV), detectors have discovered mysterious gamma-ray bursts. The longer-duration (more than a few seconds) bursts are now thought to arise from "hypernovae" associated with the collapse of rotating cores of massive stars (Section 20.2), while the shorter-duration bursts are understood to originate from the "kilonovae" associated with the merger of neutron stars (Section 20.6).

Finally, orbiting telescopes have also been used for the part of the infrared blocked by the atmosphere. Such infrared observation are particularly key to studying cool dense regions of the interstellar medium where dust absorption leads to many magnitudes of extinction in visible light; these are often regions of active star formation (Sections 12.3 and 21.3, and Chapter 22).

A full list of space-based telescopes is given at:

`https://en.wikipedia.org/wiki/List_of_space_telescopes.`

13.5   Polarimetry: Detecting Linear and Circular Polarization

So far we have focused on the intensity and flux of light, and how it can change with wavelength. But light is an *electromagnetic wave*, in which the electric and magnetic fields oscillate in the directions perpendicular to each other, and to the direction of wave propagation. As illustrated in Figure 13.3, unpolarized light includes electric vector oscillations in both perpendicular directions. But a polarizing plate can convert this into *linear polarized* light, in which the electric vector oscillates only in a single plane. By what is known as a quarter-wave plate, such linear polarized light can then be converted into *circularly polarized* light, in which case the electric vector rotates, with either a right-handed or left-handed sense, through a circle with each period as the wave propagates.

In astrophysical sources, linear polarization can develop from scattering, e.g., by electrons, from clumps or a disk, meaning that detection of such linear polarization from an spatially unresolved source can give clues to such spatial structure.

Circular polarization most commonly arises from line emission in the presence of a magnetic field, through what is known as the Zeeman effect. This provides a key way to detect magnetic fields.

In addition to the light's intensity I, its linear polarization in the two perpendicular planes are referred to as Q and U, while its circular polarization is denoted V. The full

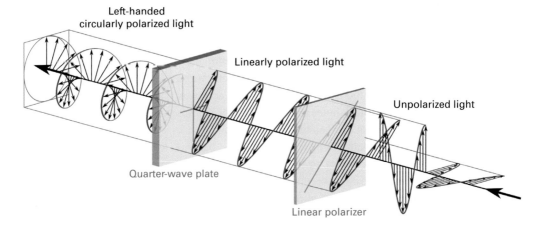

Figure 13.3 Illustration of how an unpolarized electromagnetic wave, with its electric vector oscillating in both directions perpendicular to its propagation, can be converted into a linearly polarized wave, with the electric vector in one plane, and then into circular polarization, with its electric vector rotating as the wave propagates.

set $I\,Q\,U\,V$ are called the *Stokes parameters*. Telescopes equipped with *polarimeters*, particularly when combined with spectrometers that resolve the wavelength variation, known then as *spectropolarimeters*, can thus provide key additional clues to the spatial structure of an astrophysical source.

Moreover, the Stokes V, which measures the difference in right-handed versus left-handed circular polarization, is a direct measure of the strength of the magnetic field component, along the observer's line of sight, $V \sim B_{los}$. An example of this is given in the "magnetogram" image of the Sun in the second panel from the left in the top row of Figure 14.2, for which the white and black areas denote different signs of Stokes V, reflecting the different polarity of the surface magnetic fields in the two neighboring sunspots pairs.

13.6   Questions and Exercises

Quick Questions

1. What is the limiting magnitude for a naked-eye viewing through a 5 m telescope?
2. Estimate the angular resolution of a 5 m optical telescope in space.
3. What is the diameter D of a single telescope with the same collecting area of the four 8 m-diameter VLT telescopes in Chile?
4. What is the angular resolution of the Hubble Space Telescope at a wavelength of 200 nm?
5. Two infrared sources in the Orion nebula, which has a distance $d = 500\,pc$, are separated by just $s = 0.1\,pc$. What diameter telescope would be needed to resolve them at a wavelength $\lambda = 100\,\mu m$?

Exercises

1. *Very Large Array*

   The Very Large Array in New Mexico consists of 27 radio telescopes that are each 25 m in diameter, which can be spread out over landscape of up to $a = 36\,km$ across.

   a. What is the total collecting area of the array, in m^2?

   b. What would be the diameter D_1 (in m) of a single telescope of the same area?

   c. For its range of operating frequencies 1–30 GHz, what is the associated wavelength range in cm?

   d. Then for $D = 36\,km$, work out the associated range in angular resolution α, in arcsec.

   e. How much finer is this than the angular resolution of a single telescope of the same collecting area as the full array?

2. *Limiting magnitude*

   The human eye has an integration time $t_{int} \approx 0.1\,s$, and a photon detection efficiency $\epsilon \approx 0.1$. Generalize Eq. (13.2) to estimate the m_{lim} for a telescope detector with higher values of t_{int} and ϵ.

3. *Detecting an IR-emitting cloud*

   Consider an interstellar cloud with diameter $D = 1$ pc at a distance $d = 500$ pc, which radiates like a blackbody of temperature $T = 50$ K.

   a. What is the flux F (in W/m^2) that reaches Earth?
   b. What is the wavelength λ_{max} of peak flux, in μm?
   c. For a 1 m infrared (IR) telescope observing at wavelength $\lambda = 100$ μm, what is total detected energy rate E_λ, in J/s/μm?
   d. Now suppose the telescope mirror itself radiates like a blackbody of temperature $T = 300$ K, but with an efficiency of just 0.01 percent. How does its energy emission rate compare to the answer to part c?
   e. How does this change if we cool the mirror to a temperature of 30 K?

14 Our Sun

Thus far, our discussion of stellar properties has mainly used our Sun as a benchmark for key overall quantities, such as surface temperature, radius, mass, and luminosity. But of course the close proximity of the Sun, and its extreme apparent brightness, makes it by far the most important star for own lives here on Earth. Other stars are so far away that even to our most powerful telescopes they appear as mere points of light, from which we can only measure the overall flux, or apparent brightness. But the Sun is close enough that we can resolve its *surface brightness*, or intensity, across its angular diameter of about 0.5 degree. When its extreme brightness is suitably filtered by a dark lens, it appears to our eyes as a generally featureless disk. But even with his small, primitive telescope, Galileo was able to discover darkened blemishes we now call sunspots, and so disprove the classical ideal of the Sun as a perfect, heavenly sphere.

In modern times we have access to powerful telescopes, both on the ground and in space, that observe and monitor the Sun over a wide range of wavelength bands. These vividly demonstrate that the Sun is, in fact, highly structured and variable over a wide range of spatial and temporal scales, and so provide a sobering reality check on our own simple idealizations of stars as being constant, featureless, spherically symmetric balls of gas.

The lower panel of Figure 14.1 provides a schematic summary of the various layers and features of the Sun's interior and atmosphere. From the Sun's very hot ($\sim$15 MK) nuclear-burning core, the temperature declines to the 5800 K of its visible surface, known as the *photosphere*. But a key surprise from such solar observations is that this outward decline in temperature becomes *reversed* in the Sun's outer atmosphere. As illustrated in the upper panel of Figure 14.1, from the 5800 K effective temperature at the base of the visible photosphere, the temperature declines to minimum of about $\sim$4200 K, but then rises, first gradually to several times 10^4 K through the solar *chromosphere*, and then suddenly jumping over a narrow *transition region* to more than a million kelvin in the solar *corona*.

14.1 Imaging the Solar Disk

Figure 14.2 shows images of the solar disk made by NASA's orbiting Solar Dynamics Observatory (SDO) in 13 different wavebands, chosen to highlight these different

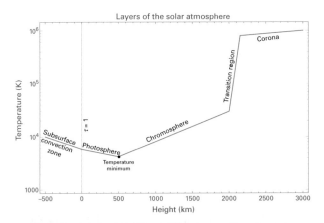

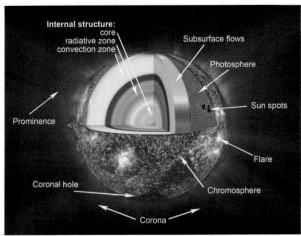

Figure 14.1 Upper panel: Schematic of temperature variation with height in the Sun's atmosphere, identifying its key layers. Observations at visible wavelengths penetrate to a layer where the optical depth $\tau = 1$, which defines the base of the *photosphere*, with an effective temperature of about 5800 K. From the very hot (MK) core the temperature declines, through the subsurface convection zone and the photosphere, but reaches a minimum value ($\sim$ 4200 K). It then rises gradually through the *chromosphere*, and then suddenly jumps over a narrow *transition region* to more than a million kelvin (MK) in the solar corona. Lower panel: Schematic summary of key regions and features of the solar atmosphere, along with interior cutout showing the Sun's nuclear-burning core, intermediate radiative diffusion region, and near-surface convection zone. Image credit: NASA.

layers of the solar atmosphere, corresponding to the labeled temperatures. In the top row, the third image from the left shows the standard visual continuum, often dubbed "white-light," formed in the photosphere. As detailed in Appendix D, in Section D.2, and further illustrated in Figure D.2, the less-bright, redder intensity toward the disk edge, known as *limb darkening*, results from the vertical decline of temperature through this photospheric layer.

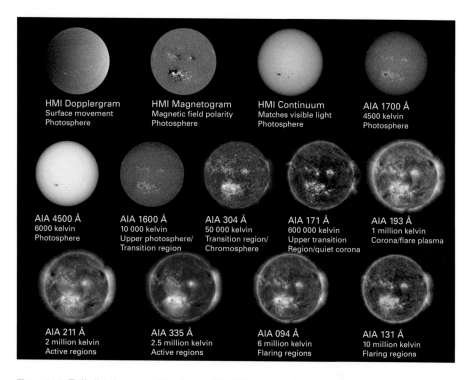

Figure 14.2 Full-disk images of the Sun at 13 different wavebands, made by NASA's Solar Dynamics Observatory (SDO). This includes images from both from the Advanced Imaging Assembly (AIA), which helps show how solar material moves around the Sun's atmosphere, and the Helioseismic and Magnetic Imager (HMI), which focuses on the movement and magnetic properties of the Sun's surface. Each wavelength was chosen to highlight a particular part of the Sun's atmosphere, from the solar photosphere, through the chromosphere, and up to the upper reaches of the corona. Credits: NASA/SDO/Goddard Space Flight Center. Further details at: www.nasa.gov/content/goddard/how-sdo-sees-the-sun.

The sunspots below and to the left of the disk center appear dark in this visual image because, as shown in the "magnetogram" just to its left,[1] these are regions of strong magnetic field, with the light-to-dark switch indicating a change in the magnetic polarity; the fields are so strong that they inhibit the convective transport of energy from below, thus making sunspots relatively cool (~3800 K), and thus darker.

But these fields also are conduits for magnetic waves and turbulence. When dissipated at higher layers they add extra mechanical heating that causes the temperature in these upper layers to rise.

In the more-opaque UV wavebands that are formed in these higher layers, the regions above sunspots, known as *active regions*, are thus actually *brighter* than the surrounding areas. For example, in the central panel of the middle row, which is tuned

---

[1] Such magnetograms detect the circular polarization of light induced by magnetic fields on the solar surface, as described in Section 13.5.

to 304 Å (30.4 nm) emission from ionized helium at temperatures of 50 000 K in the upper chromosphere, the active regions are bright, though there is still emission over the entire solar disk. But as one moves to the *far* UV and X-ray diagnostics (right middle and bottom row) that are formed at the MK temperatures of the corona, the contrast becomes greater, with some nearly dark regions that have little or no emission, known as *coronal holes*.

14.2   Corona and Solar Wind

Although very hot, the corona has a very low density, even above active regions. At visual wavelengths it is thus nearly transparent, and so generally hard to see. Fortunately, by an amazing coincidence, Earth's Moon has nearly the same angular size as the Sun, and so, in rare and brief instances, there occurs a *solar eclipse*, during which the Moon just covers up the bright solar disk. As shown in Figure 14.3, this allows us to see the corona as visible solar light scattered by electrons in the corona's tenuous, but highly ionized, gas. In the right panel, the rim of red light comes from the chromosphere,[2] via the magnetic suspension of hot gas in active regions, leading to hydrogen Balmer-α ($n = 3$ to $n = 2$; see Appendix A) emission at the red wavelength 6563 Å (656.2 nm).

The magnetic fields from these active regions rise up into the corona, forming *closed magnetic loops* that connect footpoints of opposite magnetic polarity on the surface (see Figure 14.4). In fact, the corona is so hot that the Sun's gravity cannot, by itself, keep the gas bound against a pressure-driven outward expansion known as the

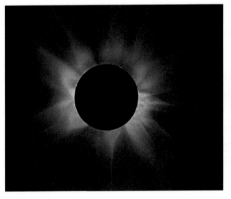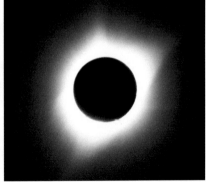

Figure 14.3 Left: Image of 1980 eclipse made with special radial density filter to show extensive structure of the solar corona during a time of maximum magnetic activity. Credit: NCAR's High Altitude Observatory. Right: Basic camera image of 2017 eclipse, showing the white corona and the red chromospheric emission around the solar limb. Credit: Alan Ford.

[2] This red color led to the name "chromosphere," from the Greek *chroma* for color.

Figure 14.4 Gas filaments trace out closed magnetic loops extending above the solar limb, as imaged in the UV by NASA's Transition Region and Coronal Explorer (TRACE) satellite. Credit: NASA/TRACE/LMSAL.

solar wind. But in regions with closed magnetic loops, the magnetic field tension holds the gas back against this expansion, allowing such regions to keep a high pressure and density, and thus making them more visible in both white-light and X-ray signatures.

Coronal holes arise in *open* field regions between such closed loops, allowing the gas to escape into the outward solar wind expansion. This gives a lower coronal density and relatively low brightness both in scattered white-light and X-ray emission. The coronal magnetic field is thus the key cause of the coronal structure seen in the left panel of Figure 14.3.

The radial streamers[3] at the tops of the coronal loops show that wind expansion wins out in the outer corona, effectively pulling open the closed field lines there. The resulting solar wind expands outward, past the Earth and even all the other planets, extending to distances greater than 100 au, until it is finally stopped by running into the local interstellar medium. The full region within this wind-termination boundary is referred to as the *heliosphere*.

As illustrated in Figure 24.2, the magnetosphere formed by the Earth's own magnetic field shields our planet and its atmosphere from a direct hit by the solar wind, instead just channeling any solar wind plasma toward the magnetic poles, where interaction with Earth's atmosphere forms the aurorae, also known as the northern and southern lights. In contrast, the lack of a strong field on Mars has allowed the solar wind to gradually erode its now much thinner atmosphere. As discussed in Section 24.3, this can affect the habitability of extra-solar planets around cool stars with coronal winds.

---

[3] Sometimes referred to as "helmet streamers," due their resemblance to German World War I army helmets.

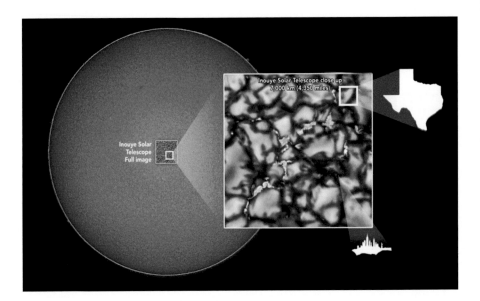

Figure 14.5 Illustration of how the DKIST telescope allows us to zoom in to image structure on the solar surface down to a resolution of ~0.1 arcsec, or ~70 km, only about twice the size of Manhattan. The irregular granulation structures have typical size of ~2 arcsec, or ~1500 km, about the size of Texas. They represent cells of convection, with brighter center upwelling from the hotter interior, bounded by narrow lanes of cooler darker downflows. Image credit: NSO/NSF/AURA; visit www.nso.edu.

14.3   Convection as a Driver of Magnetic Structure and Activity

The angular diameter of the solar disk is

$$\alpha_\odot = \frac{2R_\odot}{\text{au}} \approx 0.01 \text{ rad} \approx 0.5 \text{ degree} \approx 1800 \text{ arcsec}. \tag{14.1}$$

This means the roughly 1 arcsec resolution limit from atmospheric seeing allows for about 1800 resolution elements across the solar disk, representing a physical size of $s \approx 2R_\odot/1800 \approx 700$ km. With special techniques to correct for this atmospheric seeing, it is possible to reach a factor 10 higher resolution, so down to 0.1 arcsec, or a physical size $s \approx 70$ km.

As illustrated in Figure 14.5, such resolution is achieved by DKIST,[4] the currently most advanced ground-based solar telescope. Its primary mirror has a diameter $D = 4$ m, which from Eq. (13.3) gives a diffraction-limit resolution <0.1 arcsec in the visible. Its site at an altitude of about 3000 m atop the Haleakala volcano on the island of Maui, Hawaii, was chosen for its relatively stable air and so good, subarcsec seeing, which with adaptive-optics correction allows resolution that approaches this diffraction limit.

---

[4] For *Daniel K. Inouye Solar Telescope*, named in honor of the Hawaii senator who championed funding for the project in the US congress.

Zooming in to a small segment of the disk, Figure 14.5 shows the Sun's *granulation* pattern, with central bright cells bounded by narrow, darker lanes. This is characteristic of a systematic gas motion called *convection*. Hotter gas in the interior wells upward in the cell centers, making them hotter and thus brighter. After this gas cools by radiation into space, it falls back downward in the narrow lanes that bound the cells, which, being cooler, also appear darker. As detailed in Section 17.3, such convection arises in the near-surface layers of relatively cool stars such as the Sun, from the blocking of radiative diffusion by the enhanced opacity associated with ionization of neutral hydrogen. An animation of the dynamical variation of this convective structure is given in www.youtube.com/watch?v=4nieF-e0OOs.

Such convection combines with the Sun's rotation to generate magnetic fields through a "rotation–convection magnetic dynamo." Although hydrogen gas in the solar atmosphere is mostly neutral, other elements lose a sufficient number of their less tightly bound electrons to make the overall gas behave like an ionized plasma, with a high electrical conductivity. Such a conducting plasma makes any magnetic field "frozen-in," or effectively stuck to the local plasma. Near and below the stellar surface, where the gas energy density dominates over that associated with the magnetic field, the stretching and compression of any embedded magnetic field acts to amplify that field. For example, in granulation convection cells, the dense material upwelling in the center tends to sweep any magnetic field lines to the cell edges, thus concentrating the field in the narrow dark lanes. The small-scale bright regions in the dark lanes in Figure 14.5 are sites of such locally concentrated magnetic field.

Above the solar surface, the rapid decrease in gas density and pressure with height means that in the upper layers of the Sun's atmosphere, in the chromosphere and extending up into surrounding corona, it is the magnetic field that dominates and channels the gas, leading to the extensive coronal structure shown in the eclipse image in Figure 14.3, and the magnetic loops seen above the solar limb in Figure 14.4.

Finally, as illustrated in Figure 14.6, larger-scale interior generation of magnetic fields occurs through the interaction of convection with the Sun's *differential rotation*. The latter refers to the fact that the Sun does not rotate as a solid body, like the Earth or other rocky planets, but instead actually has a faster angular rotation (shorter rotation period) at its equator than at higher latitudes towards its poles. As field lines

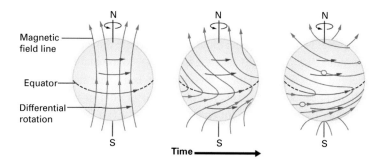

Figure 14.6 Schematic showing how differential rotation winds up and amplifies a large-scale, initially dipolar field.

are stretched azimuthally, they eventually form kinks that pop up through the solar photosphere, forming sunspot pairs with opposite magnetic polarity.

The increased strength and complexity of the field, coupled with footpoint wandering induced by convection, can lead to localized regions of *magnetic reconnection*, effectively canceling field lines of opposite polarity. The associated sudden release of magnetic energy leads to a localized *flare* in brightness in wavebands from the visible to X-ray, as shown in several panels on the bottom row of Figure 14.2.

Over time this reconnection dissipation leads to an overall decline in magnetic field strength and complexity, and so an associated decline in solar activity till this reaches a relatively quiescent minimum, whereupon the cycle restarts with winding up of the large-scale residual field by differential rotation. This is the origin of the 11-year cycle seen in, e.g., sunspot number, as well as other signatures of solar activity.

Through monitoring spectroscopic signatures of activity, including coronal X-ray emission, activity cycles have been inferred in other cool, solar-type stars, albeit with varying periods ranging from about a year to many decades. This illustrates again how the Sun provides us with a benchmark for complex structure and activity in stars that we see only as points of light, reminding us that they, too, are far more complex than our idealized steady, spherical models would suggest.

14.4   Questions and Exercises

Quick Questions

1. For solar-wind mass-loss rate $\dot{M} \approx 10^{-14}$ $M_\odot$/yr and speed $v = 450$ km/s, what is the associated kinetic energy luminosity L_{sw} (in both $L_\odot$ and W) of the solar wind?

2. The Sun's surface temperature is about $T_\odot = 5800$ K, but that of a sunspot is about $T_{spot} = 3800$ K. What is the ratio of the surface flux of a sunspot compared with the rest of the Sun, F_{spot}/F_{sun}.

3. Referring to Appendix D on radiative transfer in the atmospheres of the Sun and other stars, derive the integral solution (D.3) from the differential equation (D.1), assuming a semi-infinite atmosphere that extends to large depths $\tau \to \infty$. (Hint: First multiply Eq. (D.1) by an integrating factor $e^{-\tau/\mu}$, and use this to write the change in intensity in terms of a full differential. Then carry out the integral from the observer at $\tau = 0$ to some finite depth τ, where the intensity is taken to have a given value $I(\tau, \mu)$. Finally take the limit $\tau \to \infty$ to obtain Eq. (D.3).)

4. If thermal emission from the Planck function is a linear function of radial optical depth $B(\tau) = a + b\tau$, explicitly do the integration in Eq. (D.3) to derive the Eddington–Barbier relation for emergent intensity $I(\mu,0) = B(\tau = \mu)$.

5. Now suppose the Planck function is a quadratic function of radial optical depth, $B(\tau) = a + b\tau + c\tau^2$. Derive the associated variation of emergent intensity $I(\mu,0)$, and determine the conditions for which the limb would be brighter than the disk center, i.e., $I(0,0) > I(1,0)$. This is known as "limb brightening." For $a = b = 1$ and $I(0,0) = 2I(1,0)$ derive the value of c, and plot $I(\mu,0)$ versus μ.

Exercises

1. *Sunspot reduction in solar emission*
   Sunspots have a temperature of about 3800 K and can be as large as 50 000 km in diameter.
   a. What fraction a of the Sun's total surface area does one such sunspot cover?
   b. From the blackbody law (Eq. (5.1)), what is the reduction factor f in the radiative flux from a sunspot compared to the regular Sun with temperature $T_{\text{eff}} = 5800$ K?
   c. From parts a and b, what is the fractional reduction s in the solar luminosity from each such sunspot? (Ignore here the fact that such sunspot reduction is partly compensated by regions of extra brightness.)
   d. If the Earth were also a blackbody in absorbing and reradiating the incoming solar flux, estimate the associated reduction in Earth's equilibrium temperature, ΔT_{eq} (in K) from a single such spot at the center of the Sun's disk. (Hint: See Section 23.5.)

2. *Solar coronal density*
   The Sun and full Moon have apparent magnitudes of respectively -26.74 and -12.74. Coincidentally, during a total eclipse, the Sun's corona has roughly the same apparent brightness as a full Moon.
   a. Use this to estimate what fraction f of the Sun's luminosity is scattered by electrons in the solar corona.
   b. Associating this fraction with the corona's optical thickness τ_e to electron scattering, compute the associated electron column density N_e of the solar corona, in cm^{-2}. (Hint: See Section C.1 in Appendix C for the Thomson cross section for electron scattering.)
   c. For a simple model where the electron volume density n_e in the corona is roughly constant over a solar radius $R_\odot$ above the solar surface (and zero above this), compute n_e in cm^{-3}.

3. *Solar wind*
   Interplanetary spacecraft at a Sun-distance $r = 1$ au measure typical solar-wind proton density $n_p \approx 7$ cm^{-3} and speed $v_p \approx 450$ km/s.
   a. Assuming a spherically symmetric solar wind expansion, what is the associated solar wind mass loss rate $\dot{M}$, in $M_\odot$/yr?
   b. How does this compare with the mass loss associated with the Sun's luminosity, $\dot{M}_{\text{lum}} = L_\odot/c^2$?
   c. For a characteristic proton density $n_p = 10^8$ cm^{-3} at the $r \approx R_\odot$ at the base of the solar corona, use mass flux conservation to estimate the flow speed at this coronal base, $v_p(R_\odot)$.
   d. Over the Sun's 10 Gyr main sequence lifetime, what fraction f of the Sun's mass will be lost via the solar wind?

Part II

Stellar Structure and Evolution

15 Hydrostatic Balance between Pressure and Gravity

We have seen in Part I how a star's color or peak wavelength $\lambda_{\max}$ indicates its characteristic temperature near the stellar surface. But what about the temperature in the star's deep interior? Intuitively, we expect this to be much higher than at the surface, but under what conditions does it become hot enough to allow for nuclear fusion to power the star's luminosity? And how does it scale quantitatively with the overall stellar properties, such as mass M, radius R, and perhaps luminosity L?

To answer these questions, let us identify two distinct considerations for our intuition that the interior temperature should be much higher than at the surface.

The first we might characterize as the "blanketing" by the overlying layers, which traps any energy generated in the interior, much as a blanket in bed traps our body heat, keeping our skin temperature at a comfortable warmth, instead of the relative chilliness of having it exposed to open air. In this picture, the equilibrium interior temperature depends on the rate of energy generation (from metabolism for our bodies, or nuclear fusion for stars) and the "insulation thickness" of the overlying layer of material to the surface (given by the optical depth; see Chapter 16.)

But distinct from this consideration of the *transport of energy* from the interior, there is for a star a dynamical requirement for *force* or *momentum* balance, to keep the star supported against the inward pull of its own self-gravity. Since stars are gaseous, without the tensile strength of a solid body, this gravitational support is supplied by increased internal gas pressure P, allowing the star to remain in a static equilibrium. This high gas pressure arises from a combination of high density and high temperature. As detailed in Section 15.3, this allows us to determine a characteristic interior temperature, through a further application of the Virial Theorem for bound systems that was briefly discussed for bound orbits in Section 7.4.

15.1   Hydrostatic Equilibrium

To quantify this gravitational equilibrium for a static star, consider, as illustrated in Figure 15.1, a thin radial segment of thickness dr with local density ρ and downward gravitational acceleration g. The mass-per-unit-area of this layer is $dm = \rho dr$, with corresponding weight-per-unit-area $g\,dm$. To support this weight, the gas pressure at

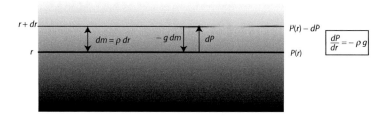

Figure 15.1 Illustration of the radial decline of gas pressure P due to local mass density ρ and downward gravity g.

the lower end of this layer must be higher by the amount $|dP| = g\,dm = \rho g\,dr$ than the upper end, implying

$$\frac{dP}{dr} = -\rho g,\tag{15.1}$$

a condition known as *hydrostatic equilibrium.*

For an *ideal gas*, the pressure depends on the product of the number density $n = \rho/\bar{\mu}$ and temperature T,

$$P = nkT = \rho\frac{kT}{\bar{\mu}} \equiv \rho c_s^2,\tag{15.2}$$

where $k = 1.38 \times 10^{-16}$ erg/K is Boltzmann's constant, and $\bar{\mu}$ is the average mass – the "mean molecular weight" – of all particles (i.e., both ions and electrons) in the gas; the final equation defines the isothermal[1] sound speed, $c_s \equiv \sqrt{kT/\bar{\mu}}$.

Because the electron mass m_e is negligibly small compared with the nearly equal proton or neutron mass, $m_p \approx m_n$, the *fully ionized* molecular weight for any given element of atomic number Z_n and atomic weight A is just $\mu = m_p A/(Z_n + 1)$. For a gas mixture with mass fraction X, Y, and Z for H, He, and metals, the overall mean molecular weight is then obtained by a weighted average of the *inverses* ($m_p/\mu = (Z_n + 1)/A$) of the individual components (as in a parallel circuit), yielding

$$\bar{\mu} = \frac{m_p}{2X + 3Y/4 + Z/2} \approx 0.6 m_p \equiv \bar{\mu}_\odot,\tag{15.3}$$

where the last equality is for the solar case with $X = 0.72$, $Y = 0.26$, and $Z = 0.02$. More generally, for fully ionized gases the proton-mass-scaled molecular weight $\bar{\mu}/m_p$ can range from 1/2 for pure H ($X = 1$), to 4/3 for pure He ($Y = 1$), to a maximum of 2 for pure heavy metals ($Z = 1$).

---

[1] This speed, which was first derived by Newton, would only be the speed of sound if the gas remained strictly constant temperature (isothermal). In practice, the temperature fluctuations associated with the gas compressions make the actual "adiabatic" speed of sound slightly higher, by a factor $\sqrt{\gamma}$, where γ is the ratio of specific heats (5/3 for a monatomic gas).

15.2   Pressure Scale Height and Thinness of Surface Layer

The ratio of Eq. (15.2) to Eq. (15.1) defines a characteristic *pressure scale height,*

$$H \equiv \frac{P}{|dP/dr|} = \frac{kT}{\bar{\mu}g} = \frac{c_s^2}{g}, \tag{15.4}$$

where the absolute value of the pressure gradient dP/dr (which itself is negative) ensures the scale height is positive.

At the stellar surface radius $r = R$, where the gravity and temperature approach their fixed surface values $g_* = GM/R^2$ and $T = T_*$, the scale height becomes quite small, typically only a tiny fraction of the stellar radius,

$$\frac{H}{R} = \frac{kT_*/\bar{\mu}}{GM/R} = \frac{2c_{s*}^2}{V_{\text{esc}}^2} \approx 0.0005 \, \frac{T_*/T_\odot}{\bar{\mu}/\bar{\mu}_\odot} \frac{R/R_\odot}{M/M_\odot}, \tag{15.5}$$

where V_{esc} is the escape speed introduced in Section 7.2. For the solar atmosphere, the sound speed is $c_{s*} \approx 9$ km/s, about 1/60th of the surface escape speed $V_{\text{esc}} = 620$ km/s.

If we further idealize a stellar atmosphere as being roughly isothermal, i.e. with a nearly constant temperature $T \approx T_*$, then, since the gravity is also effectively fixed at the surface value, we see that the scale height also becomes constant. This makes it easy to integrate the hydrostatic equilibrium equation (15.1), thus giving the variation of density and pressure in terms of a simple exponential stratification with height $z \equiv r - R$,

$$\frac{P(z)}{P_*} = \frac{\rho(z)}{\rho_*} = e^{-z/H}, \tag{15.6}$$

where the asterisk subscripts denote values at some some surface layer where $z = 0$ (or $r = R$). In practice the temperature variations in an atmosphere are gradual enough that, quite generally, both pressure and density very nearly follow such an exponential stratification.

The results in this section actually apply to *any* gravitationally bound atmosphere, not only for stars but also for planets, including the Earth, with similarly small characteristic values for the ratio H/R. This is the basic reason that the Earth's atmosphere is confined to such a narrow layer around its solid surface, meaning that at an altitude of just a couple of hundred kilometers it is nearly a vacuum, so tenuous that it imparts only a weak drag on orbiting satellites.

For stars or gaseous giant planets without a solid surface, it means that the dense, opaque regions have only a similarly narrow transition to the fully transparent upper layers, thus giving them a similarly sharp visual edge as a solid body. For stars it means that models of the escape of interior radiation through this narrow atmospheric layer can essentially ignore the stellar radius, allowing the emergent spectrum to be well described by a planar atmospheric model fixed by just two parameters – surface temperature and gravity – and not dependent on the actual stellar radius.

15.3   Hydrostatic Balance in the Stellar Interior and the Virial Temperature

This hydrostatic balance must also apply in the stellar interior, but now both the temperature and gravity have a strong spatial variation. At any given interior radius r, the local gravitational acceleration depends only on the mass within that radius,

$$M(r) \equiv 4\pi \int_0^r \rho(r')r'^2 \, dr'. \tag{15.7}$$

This thus requires the hydrostatic equilibrium equation to be written in the somewhat more general form,

$$\boxed{\frac{dP}{dr} = -\rho(r)\frac{GM(r)}{r^2}.} \tag{15.8}$$

This represents one of the key equations for stellar structure.

The implications of hydrostatic equilibrium for the hot interior of stars are quite different from the steep exponential pressure drop near the surface; indeed they allow us now to derive a remarkably simple scaling relation for a characteristic interior temperature T_{int}.

For this, consider the associated interior pressure P_{int} at the center of the star ($r = 0$); to drop from this high central pressure to the near-zero pressure at the surface, the pressure gradient averaged over the whole star must be $|dP/dr| \sim P_{int}/R$. We can similarly characterize the gravitational attraction in terms of the surface gravity $g_* = GM/R^2$ times an interior density that scales as $\rho_{int} \sim P_{int}\bar{\mu}/kT_{int}$. Applying these in the basic definition of scale height (Eq. (15.4)), we find that, for the interior, $H \approx R$, which in turn implies for this characteristic stellar interior temperature,

$$\boxed{T_{int} \approx \frac{GM\bar{\mu}}{kR} \approx 14 \times 10^6 \text{ K} \frac{M/M_\odot}{R/R_\odot}.} \tag{15.9}$$

Thus, while surface temperatures of stars are typically a few thousand kelvin, we see that their interior temperatures are typically of order 10 *million* kelvin! As discussed in Chapter 18, this is indeed near the temperature needed for nuclear fusion of hydrogen into helium in the stellar core.

This close connection between thermal energy of the interior ($\sim kT$) to the star's gravitational binding energy ($\sim GM\bar{\mu}/R$) is really just another example of the Virial Theorem for gravitationally bound systems, as discussed in Section 7.4 for the case of bound orbits. The temperature is effectively a measure of the average kinetic energy associated with the random thermal motion of the particles in the gas. Thermal energy is thus just a specific form of kinetic energy, and the Virial Theorem tells us that the average kinetic energy in a bound system equals one-half the magnitude of the gravitational binding energy.

15.4 Questions and Exercises

Quick Questions

1. If we double the Sun's radius, what happens to its surface gravity?
2. If we double the Sun's mass, what happens to its surface gravity?
3. If we double the Sun's surface gravity, what happens to its surface scale height H?
4. If we double the Sun's surface temperature, what happens to its surface scale height H?
5. If we double the Sun's radius, what happens to its core temperature?
6. If we double the Sun's mass, what happens to its core temperature?

Exercises

1. *Escape speed and central temperature*
   Compute the escape speed (in km/s) from stars with:
   a. $M = M_\odot$ and $R = 4R_\odot$
   b. $M = 4M_\odot$ and $R = R_\odot$
   c. $M = 4M_\odot$ and $R = 4R_\odot$.
   d. For these stars, estimate the associated central temperatures.

2. *Constant-density star*
   a. For a constant-density star of mass M and radius R, compute the mass within radius r.
   b. Compute next the radial variation of gravity $g(r)$.
   c. Finally, show by explicit integration of the hydrostatic equilibrium equation (Eq. (15.8)) that the core temperature is half the value given by Eq. (15.9).

3. *Helium stars*
   Normal stars are made mostly of hydrogen, but consider now "helium stars," in which all the hydrogen has been somehow depleted, or converted into helium.
   a. For a fully ionized mixture in which the hydrogen, helium, and metal mass fractions are now $X = 0$, $Y = 0.98$, and $Z = 0.02$, compute the mean molecular weight $\bar{\mu}$ for such a helium star.
   b. For a temperature $T = 10\,000$ K, compute now the associated sound speed c_s, and compare it with the value for a star with solar abundances at the same temperature.
   c. For a helium star with the solar mass and radius but $T = 10\,000$ K, what is the scale height H, both in km and fractions of the stellar radius?
   d. For this star, estimate now the interior temperature T_{int}, and compare it to the interior temperature for a star with solar abundances.

4. *Scale height in Earth's atmosphere*
   a. The Earth's atmosphere is mostly diatomic nitrogen, with molecular weight $\mu \approx 28m_p$. For a typical temperature on a spring day ($\sim 50\,°F$), compute the isothermal sound speed, c_s, in km/s, and as a ratio to the orbital speed in low-Earth orbit, $v_{orb} = 7.9$ km/s.

b. Use this and Earth's surface gravity to compute the atmospheric scale height H for the Earth (in km), and its ratio to Earth's radius, H/R_e. How does the latter compare with c_s/v_{orb}?

c. The pressure at sea level is defined as 1 atmosphere (atm). Ignoring any temperature change of the atmosphere, estimate the pressure (in atm) at a typical altitude $h = 300$ km for an orbiting satellite.

d. A satellite in circular orbit at this altitude of $h = 300$ km will typically stay in orbit for about decade. If the temperature of the remaining gas at this height is twice that of the Earth's surface, estimate how much higher the orbital height would have to be to double this orbital lifetime.

16 Transport of Radiation from Interior to Surface

16.1 Random Walk of Photon Diffusion from Stellar Core to Surface

Let us now turn to the "blanketing" effect of the star's material in trapping the heat and radiation of the interior. Within a star, the absorption of light by stellar material is counteracted by thermal emission. As illustrated in Figure 16.1, radiation generated in the deep interior of a star undergoes a diffusion between multiple encounters with the stellar material before it can escape freely into space from the stellar surface. The number of mean-free-paths ℓ from the center at $r = 0$ to the surface at radius $r = R$ now defines the central optical depth

$$\tau_c = \int_0^R \frac{dr'}{\ell} = \int_0^R \kappa \rho \, dr'. \tag{16.1}$$

As discussed in Appendix C, the opacity in stellar interiors typically has a CGS value near unity, $\kappa \approx 1 \, \text{cm}^2/\text{g}$, with a minimum set by the value for Thomson scattering by free electrons, $\kappa_e \approx 0.34 \, \text{cm}^2/\text{g}$. We can then estimate a typical value of this interior optical depth simply by taking the density to be roughly characterized by its volume average $\bar{\rho} = M/(4\pi R^3/3)$, where again M and R are the stellar mass and radius. For the Sun this works out to give $\bar{\rho}_\odot \approx 1.4 \, \text{g/cm}^3$, i.e., just above the density of water. (The Sun would not quite float in your bathtub.) Since the opacity is also near unity in CGS units, the average mean-free-path for scattering in the Sun is just $\bar{\ell}_\odot \approx 0.7 \, \text{cm}$.

In the core of the Sun, the density is typically 100 times higher than this mean value, so the core mean-free-path is a factor 100 smaller, i.e., $\ell_{\text{core}} \approx 0.07 \, \text{mm}$! But either way, the mean-free-path is much, much smaller than the solar radius $R_\odot \approx 700\,000 \, \text{km} = 7 \times 10^{10} \, \text{cm}$. This implies the optical depth from the center to surface is truly enormous, with a typical value

$$\tau_c \approx \frac{R_\odot}{\bar{\ell}_\odot} \approx 10^{11}. \tag{16.2}$$

The total number of scatterings needed to diffuse from the center to the surface can then be estimated from a basic "random walk" argument. The simple one-dimensional (1D) version states that after N left/right random steps of unit length, the root-mean-square (rms) distance from the origin is $\sqrt{N}$. For the three-dimensional (3D) case of stellar diffusion, this rms number of unit steps can be roughly associated with the total

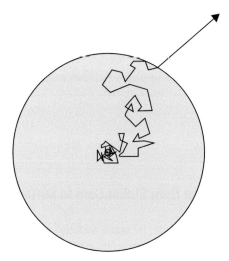

Figure 16.1 Illustration of the random-walk diffusion of photons from the core to surface of a star.

number of mean-free-paths between the core and surface, i.e., $\sqrt{N} \approx \tau$. This implies that photons created in the core of the Sun need to scatter a total of $N \approx \tau^2 \approx 10^{22}$ times to reach the surface!

In traveling from the Sun's center to its surface, the *net* distance is just the Sun's radius $R_\odot$; but the *cumulative* path length traveled is much longer, $\ell_{\text{tot}} \approx N\bar{\ell}_\odot \approx \tau^2\bar{\ell}_\odot \approx \tau R_\odot$. For photons traveling at the speed of light $c = 3 \times 10^{10}$cm/s, the total time for photons to diffuse from the center to the surface is thus

$$t_{\text{diff}} = \tau^2 \frac{\bar{\ell}_\odot}{c} \approx \tau \frac{R_\odot}{c} \approx 10^{11} \times 2.3\,\text{s} \approx 7000\,\text{yr}, \tag{16.3}$$

where, for the last evaluation, it is handy to recall again that $1\,\text{yr} \approx \pi \times 10^7$ s.

Once the photons reach the surface, they can escape the star and travel unimpeded through space, taking, for example, only a modest time $t_{\text{Earth}} = \text{au}/c \approx 8$ min to cross the 1 au ($\approx 215 R_\odot$) distance from the Sun to the Earth.

A stellar atmospheric surface thus marks a quite distinct boundary between the interior and free space. From deep within the interior, the stellar radiation field would appear nearly isotropic (same in all directions), with only a small asymmetry (of order $1/\tau$) between upward and downward photons. In contrast, near the surface, this radiation becomes distinctly anisotropic, emerging upward from the surface below, but with no radiation coming downward from empty space above.

16.2   Diffusion Approximation at Depth

This picture of photons undergoing a random walk through the stellar interior can be formalized in terms of a *diffusion* model for radiation transport in the interior.

Appendix D presents a discussion of the transition from diffusion to free-streaming that occurs in the narrow region near the stellar surface, known as the "stellar atmosphere." This is described by the *equation of radiative transfer*, given by Eq. (D.1), with Eq. (D.2) now defining the vertical optical depth $\tau(r)$ from a given radius r to an external observer at $r \to \infty$.

But in the deep interior layers within a star, i.e., with large optical depths $\tau \gg 1$, the trapping of the radiation makes the intensity I nearly isotropic and near the local Planck function B. Applying this to the derivative term in Eq. (D.1) and solving for I gives a "diffusion approximation" for the local intensity,

$$I(\mu, \tau) \approx B(\tau) + \mu \frac{dB}{d\tau},$$

(16.4)

where, as defined in Figure D.1, μ is the cosine of the angle between the ray and the vertical direction, so that $\mu = +1$ is directly upward, and $\mu = -1$ is directly downward.

Since $dB/d\tau$ is of order B/τ, we can see that the second term is much smaller, by a factor $\sim 1/\tau \ll 1$, than the first, leading-order term. Recall that both the specific intensity I and the Planck function B have the same units as a surface brightness, i.e., energy/area/time/solid angle.

The local *net upward flux* F (energy/area/time) is computed by weighting the intensity by the direction cosine μ and then integrating over a solid angle,

$$F \equiv \oint I \mu \, d\Omega = 2\pi \int_{-1}^{+1} \left(\mu B(\tau) + \mu^2 \frac{dB}{d\tau} \right) d\mu.$$

(16.5)

Since the leading-order term with $B(\tau)$ is odd over the range $-1 < \mu < 1$, it vanishes upon integration, giving

$$F \approx \frac{4\pi}{3} \frac{dB}{d\tau}.$$

(16.6)

Again recalling that $B = \sigma_{sb} T^4/\pi$, and noting the optical depth changes with radius r as $d\tau = -\kappa \rho dr$, we can alternatively write the flux as a function of the local temperature gradient,

$$F(r) = -\left[\frac{4\pi}{3\kappa\rho} \frac{\partial B}{\partial T} \right] \frac{dT}{dr} = -\left[\frac{16\sigma_{sb}}{3\kappa\rho} T^3 \right] \frac{dT}{dr}$$

(16.7)

The terms in square bracket can be thought of as a *radiative conductivity*, which we note increases with the cube of the temperature, T^3, but depends inversely on opacity and density, $1/\kappa\rho$.

16.3 Atmospheric Variation of Temperature with Optical Depth

A star's luminosity L is generated in a very hot, dense central core. Outside this core, at any stellar envelope radius r, the local radiative flux scales as $F = L/4\pi r^2$, which at the stellar radius R takes the fixed surface value

$$F_* = \frac{L}{4\pi R^2} \equiv \sigma_{sb} T_{eff}^4 \tag{16.8}$$

The latter equality defines a star's *effective* temperature, T_{eff}, as the blackbody temperature that would give an inferred surface flux, F_*.

Since in such a surface layer F_* is independent of τ, Eq. (16.6) can be trivially integrated in this layer to give

$$\frac{4\pi}{3} B(\tau) = F_*\tau + C = \sigma_{sb} T_{eff}^4 \tau + C, \tag{16.9}$$

where C is an integration constant. Recalling also from Eq. (5.1) that $\pi B = \sigma_{sb} T^4$, we can convert Eq. (16.9) into an explicit expression for the variation of temperature with optical depth,

$$\boxed{T^4(\tau) = \frac{3}{4} T_{eff}^4 \left[\tau + 2/3\right],} \tag{16.10}$$

wherein, in light of the result in Section D.2 in Appendix D, we have taken the integration constant such that $T(\tau = 2/3) = T_{eff}$.

Together with Eq. (15.1) for hydrostatic equilibrium, Eq. (16.7) for radiative diffusion determines the fundamental structure of the stellar interior. In Chapter 17, these will be used to explain the underling physics behind the main-sequence, mass–luminosity relation $L \sim M^3$, which, as discussed in Section 10.4, was found empirically from observations of binary systems with known parallax distances (see Figure 10.4).

16.4   Questions and Exercises

Quick Questions

1. If we flip a coin repeatedly, about how many flips will it take, on average, before the number of heads is 100 more than the number of tails?

2. If a star has twice the mass and twice the radius of the Sun, approximately what is the optical depth from its center to surface?

3. If we double the Sun's mass, about how long would it take light to diffuse from the center to the surface?

4. If we double the Sun's radius, approximately how long would it take light to diffuse from the center to the surface?

5. At approximately what optical depth τ is the temperature of a star 100 times its effective temperature T_{eff}?

Exercises

1. *Temperature versus optical depth*
   a. At what optical depth τ does the local temperature T in a stellar atmosphere equal the stellar effective temperature T_{eff}?
   b. At about what optical depth τ does the local temperature $T = 10 T_{eff}$?
   c. In terms of the star's density scale height H, approximately what is the physical distance between the height for parts a and b?

2. *Surface scale height and density*
   a. Near the Sun's surface where the temperature is at the effective temperature $T = T_* = T_{eff} \approx 5800\,\text{K}$, compute the scale height H (in km).
   b. Using the fact that the mean-free-path $\ell \approx H$ near this surface, compute the mass density ρ (in g/cm^3) assuming the opacity is equal to the electron scattering value given in Section C.1 in Appendix C, i.e., $\kappa_e = 0.34\,\text{cm}^2/\text{g}$.

3. *Uniform, planar, radiating slab*
   Consider a uniform planar slab that is infinite in both horizontal directions (x, y) but has a total thickness Z in the vertical direction z. Suppose the uniform mass density ρ comes from purely absorbing particles of mass m and radius r. Write expressions for the:
   a. particle number density n;
   b. particle absorption cross section σ;
   c. opacity of the medium κ;
   d. photon mean-free-path ℓ;
   e. total slab (vertical) optical thickness τ.
   Next suppose a heating element located along the slab midplane at $z = Z/2$ generates an energy flux F (energy/area/time). Assuming the slab is very optically thick ($\tau \gg 1$), derive expressions for:
   f. The total path length L that photons generated at the center ($z = Z/2$) must travel before escaping out the bottom or top (at $z = 0, Z$).
   g. $T_s = T(z = 0) = T_s(z = Z)$, the temperature at the lower and upper surfaces at $z = 0, Z$.
   h. $T_c = T(z = Z/2)$, the temperature at the slab center.

17 Structure of Radiative versus Convective Stellar Envelopes

17.1 $L \sim M^3$ Relation for Hydrostatic, Radiative Stellar Envelopes

As discussed in Part I (Section 10.4), observations of binary systems indicate that main-sequence stars follow an empirical mass–luminosity relation $L \sim M^3$. The physical basis for this can be understood by considering the two basic relations of stellar structure, namely hydrostatic equilibrium and radiative diffusion, as given in Eqs. (15.8) and (16.7).

As in the virial scaling for internal temperature given in Section 15.3, we can use a single point evaluation of the hydrostatic pressure gradient to derive a scaling between interior temperature T, stellar radius R and mass M, and molecular weight μ,

$$\frac{dP}{dr} = -\rho \frac{GM_r}{r^2}$$

$$\rho \frac{T}{\mu R} \sim \rho \frac{M}{R^2}$$

$$T R \sim M \mu. \tag{17.1}$$

Likewise, a single point evaluation of the temperature gradient in the radiative diffusion equation (Eq. (16.7)) gives

$$F = -\frac{16\sigma_{\text{sb}}}{3\kappa\rho} T^3 \frac{dT}{dr}$$

$$\frac{L}{R^2} \sim \frac{R^3}{\kappa M} T^3 \frac{T}{R}$$

$$L \sim \frac{(R T)^4}{\kappa M}$$

$$L \sim \frac{M^3 \mu^4}{\kappa}, \tag{17.2}$$

where the last scaling uses the hydrostatic equilibrium scaling in Eq. (17.1) to derive the basic scaling law $L \sim M^3$, assuming a fixed molecular weight μ and stellar opacity κ.

Two remarkable aspects of this derivation are that: (1) the role of the stellar *radius* *cancels*; and (2) the resulting M–L scaling does *not* depend on the details of the nuclear generation of the luminosity in the stellar core! Indeed, this scaling was understood from stellar structure analyses that were done (e.g., by Eddington,[1] and Schwarzschild) in the 1920s, long before hydrogen fusion was firmly established as a key energy source for the Sun and other main-sequence stars (e.g., by Hans Bethe, *c.* 1939).

17.2   Horizontal-Track Kelvin–Helmholtz Contraction to the Main Sequence

In fact, this simple $L \sim M^3$ scaling even applies to the final stages of *pre*-main-sequence evolution, when the core is not yet hot enough to start nuclear burning, but the envelope has become hot enough for radiative diffusion to dominate the transport of energy generated by the star's gravitational contraction. As the radius decreases over the Kelvin–Helmholtz timescale t_{KH} discussed in Section 8.2, the surface temperature increases in a way that keeps the luminosity nearly constant. The lower panel of Figure 17.1 illustrates that, on the H–R diagram, a late-phase pre-main-sequence star with a mass near the Sun or higher thus evolves along a *horizontal track* from right to left, stopping when it reaches the main sequence; this is where the core temperature is now high enough for hydrogen fusion (H-fusion) to take over in supplying the energy for the stellar luminosity, without any need for further contraction. As discussed in Chapter 18, for a given mass, a star's radius on the main sequence is just the value for which the interior temperature, as set by the Virial Theorem, is sufficiently high to allow this H-fusion in the core.

17.3   Convective Instability and Energy Transport

In practice, the transport of energy from the stellar interior toward the surface sometimes occurs through *convection* instead of radiative diffusion; this has important consequence for stellar structure and thus for the scaling of luminosity.

Convection refers to the overturning motions of the gas, much like the bubbling of boiling water on a stove. Stars become unstable to forming convection whenever the processes controlling the temperature make its spatial gradient too steep. This can occur in the nuclear burning core of massive stars, for which the specific mechanism for H-fusion, called the "CNO" cycle, gives the nuclear burning rate a steep dependence on temperature (Chapter 18 and Figure 18.1). The resulting steep temperature gradient makes the cores of such stars strongly convective.

Steep gradients, and their associated convection, can also occur in outer regions of cooler, lower-mass stars, where the cooler temperature induces recombination of ionized H (and He). The bound-free absorption (Figure C.1) by this neutral hydrogen

---

[1] Indeed, by assuming a simple polytropic relation $P \sim \rho^{1+1/n}$ between pressure and density, Eddington was able to derive an analytic solution for stellar structure for the case $n = 3$. Known as the "Eddington standard model," this gave explicit solutions for $L(M)$ that follows the $L \sim M^3$ scaling for moderate masses $M < 100 M_\odot$.

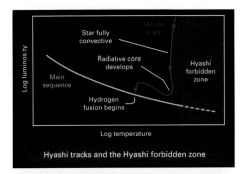

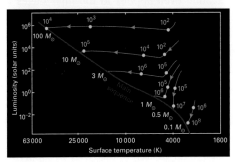

Figure 17.1 Illustration of the pre-main-sequence evolution of stars. The upper panel shows how, during the early stages of a collapsing proto-star, the interior is fully convective, causing it to evolve with decreasing luminosity at a nearly constant, relatively cool surface temperature, and so down the nearly vertical "Hayashi track" in the H–R diagram. The lower panel shows the final approach to the main sequence for stars of various masses, with green labels showing the time in years. For stars with a solar mass or above, the stellar interior becomes radiative, stopping the Hayashi track decline in luminosity. The stars then evolve horizontally and to the left on the H–R diagram, along so-called Henyey tracks, each with fixed luminosity but increasing temperature, till they reach their respective positions on the "zero-age main-sequence" or ZAMS, when the core is hot enough to ignite H-fusion.

significantly increases the local stellar opacity κ. For a fixed stellar flux $F = L/4\pi r^2$ of stellar luminosity L that needs to be transported through an interior radius r, the radiative diffusion equation (Eq. (16.7)) shows that the required radiative temperature gradient increases with such increased opacity,

$$\left.\frac{dT}{dr}\right|_{\text{rad}} = \frac{3\kappa\rho F}{16\sigma_{\text{sb}}T^3} \sim \kappa. \tag{17.3}$$

If this gradient becomes too steep, then, as illustrated in Figure 17.2, a small element of gas that is displaced slightly upward becomes less dense than its surroundings, giving it a buoyancy that causes it to rise higher still. A key assumption is that this dynamical rise of the fluid occurs much more rapidly than the rate for energy to diffuse into or out of the gas element. Processes that occur without any such energy exchange with the surroundings are called "adiabatic," with a fixed (power-law) relation of

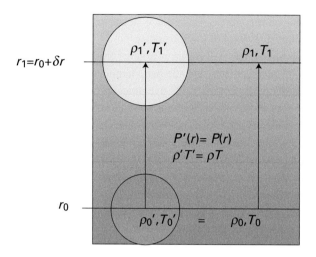

Figure 17.2 Illustration of upward displacement of a spherical fluid element in test for convective instability, which occurs when the displaced element has a lower density ρ_1' than that of its surroundings, ρ_1. Since the pressure must remain equal inside and outside the element, this requires the element to have a higher temperature, $T_1' > T_1$. Since the overall temperature gradient is negative, convection thus occurs whenever the magnitude of the atmospheric temperature gradient is steeper than the adiabatic gradient that applies for the adiabatically displaced element, i.e., $|dT/dr| > |dT/dr|_{\text{ad}}$.

pressure with density or temperature. In a hydrostatic medium with a set pressure gradient, this implies a fixed adiabatic temperature gradient $(dT/dr)_{\text{ad}}$.

Starting from an initial radius r_0 with equal density and temperature inside and outside some chosen fluid element (i.e., $\rho_0' = \rho$, $T_0' = T_0$), let us determine the density ρ_1' of that element after it is adiabatically displaced to a slightly higher radius $r_1 = r_0 + \delta r$, where the ambient density is ρ_1. Since dynamical balance requires the element and its surroundings still to have equal pressure after the displacement (i.e., $P_1' = P_1$), we have by the perfect gas law that $\rho_1' T_1' = \rho_1 T_1$. If this upward displacement $\delta r > 0$ makes the element buoyant, with lower density ρ_1' than that of its surroundings ρ_1, then, by using this constant pressure condition, we can derive the condition for the temperature gradient required for the associated convective instability,

$$\rho_1' < \rho_1; \text{ convective instability}$$

$$\frac{\rho_1'}{P_1'} < \frac{\rho_1}{P_1}$$

$$T_1 < T_1'$$

$$T_0 + \delta r (dT/dr)_{\text{rad}} < T_0 + \delta r (dT/dr)_{\text{ad}}$$

$$\left.\frac{dT}{dr}\right|_{\text{rad}} > \left.\frac{dT}{dr}\right|_{\text{ad}} \tag{17.4}$$

where, because both temperature gradients are negative, the condition in terms of absolute value requires a reversal of the inequality.

We thus see that convection will ensue whenever the magnitude of the radiative temperature gradient exceeds that of the adiabatic temperature gradient.

Convection is an inherently complex, 3D dynamical process that generally requires elaborate computer simulations to model accurately. A heuristic, semi-analytic model called "mixing length theory" has been extensively developed, but it has serious limitations, especially near the stellar surface, where the lower density and temperature can make convective transport quite inefficient. By contrast, in the dense and hot stellar interior, once convection sets in, it is so efficient at transporting energy that it keeps the local temperature gradient very close to the adiabatic value above which it is triggered.

One can thus quite generally just presume that the temperature gradient in interior convection regions is at the adiabatic value.

17.4   Hayashi Track Contraction of Fully Convective Proto-Stars

In hot stars with $T > 10\,000$ K, hydrogen remains fully ionized even to the surface; since there then is no recombination zone to increase the opacity and trigger convection, the energy transport in their stellar envelopes is by radiative diffusion. In moderately cooler stars such as the Sun (with $T_\odot \approx 6000$ K), hydrogen recombination in a zone just below the surface induces convection, which thus provides the final transport of energy toward the surface; but since the deeper interior remains ionized and thus nonconvective, the general scaling laws derived assuming radiative transport still roughly apply for such solar-type stars.

However, in much cooler stars, with surface temperatures $T \approx 3500$–4000 K, the hydrogen recombination extends deeper into the interior; this and other factors keep the opacity high enough to make the entire star convectively unstable right down to the stellar core. Because convection is so much more efficient than radiative diffusion, it can readily bring to the surface any energy generated in the interior – whether produced by gravitational contraction of the envelope, or by nuclear fusion in the core. As such, *fully convective stars* can have luminosities that greatly exceed the value implied by the $L \sim M^3$ scaling law derived in Section 17.1 (see Eq. (17.2)) for stars with radiative envelopes. As discussed in Chapter 19, this is a key factor in the high luminosity of cool giant stars that form in the post-main-sequence phases after the exhaustion of hydrogen fuel in the core.

But it also helps explain the high luminosity of the very cool, early stage of *pre-main-sequence* evolution, when gravitational contraction of a large *proto-stellar cloud* is providing the energy to make the cloud shine as a *proto-star*. Once the internal pressure generated is sufficient to establish hydrostatic equilibrium, its interior becomes fully convective, forcing the proto-star to have this characteristic surface temperature around $T \approx 3500$–4000 K.

At early stages, the proto-star's radius is very large, meaning it has a very large luminosity $L = \sigma_{sb} T^4 4\pi R^2$. As it contracts, it stays at this temperature, but the declining radius means a declining luminosity. As illustrated in the top panel of Figure 17.1, during this early phase of gravitational contraction, the proto-star thus evolves down a nearly vertical line in the H–R diagram, dubbed the "Hayashi track," after the Japanese scientist who first discovered its significance.

Once the radius reaches a level at which the luminosity is near the value predicted by the $L \sim M^3$ law, the interior switches from convective to radiative, and so the final contraction to the main sequence makes a sharp ("left") turn to a horizontal track (sometimes called the "Henyey track") with nearly constant luminosity but increasing surface temperature. The luminosity of this track is set by the stellar mass, according to the $L \sim M^3$ law derived for stars with interior energy transport by radiative diffusion. The contraction is halted when the core reaches a temperature (derived in Chapter 18; see Eq. (18.4)) for H-fusion, which then stably supplies the luminosity for the main-sequence lifetime.

As detailed in Chapter 19, once the star runs out of hydrogen fuel in its core, its *post*-main-sequence evolution effectively traces backwards along nearly the same track followed during this *pre*-main-sequence, ultimately leading to the cool, red-giant stars seen in the upper right of the H–R diagram.

17.5   Questions and Exercises

Quick Questions

1. Estimate the luminosity L (in $L_\odot$) of a radiative stellar envelope of mass $M = 30 M_\odot$.
2. What is the mass M (in $M_\odot$) of a radiative stellar envelope with luminosity $L = 64\,000 L_\odot$?

Exercises

1. *Pre-main-sequence lifetime along Henyey track*

   Consider a star that is transitioning from its fully convective structure along the Hayashi track to radiative structure along the Henyey track.

   a. For a star mass M, approximately what is the fixed luminosity L (in $L_\odot$) along this track?

   b. For this mass M, and its initial stellar temperature $T_i \approx 3000\,\text{K} \approx T_\odot/2$ at this transition, what is the initial stellar radius, R_i (in $R_\odot$)?

   c. Next, generalize Eq. (8.3) to derive an expression for the star's initial binding energy, U_i, writing this in terms of $M/M_\odot$ and the associated gravitational binding energy for Sun, $U_\odot$, as given by this Eq. (8.3).

   d. From Eq. (8.2), derive an expression for the associated scaling with mass $M/M_\odot$ of the star's Kelvin–Helmholtz timescale, $t_{KH}/t_{KM,\odot}$, relative to that of the present-day Sun.

   e. For what mass M (in $M_\odot$), would this $t_{KH} = 1$ Myr?

2. *Helium stars 2*

   This exercise builds further on Exercise 3 in Chapter 15, posed for helium stars.

   a. For a fully ionized mixture in which the hydrogen, helium, and metal mass fractions are now $X = 0$, $Y = 0.98$, and $Z = 0.02$, compute again the mean molecular weight $\bar{\mu}$ for such a helium star.

   b. From Eq. (C.2) in Appendix C, evaluate next the electron scattering opacity κ_e for such a helium star.

   c. Applying these results in the scaling (Eq. (17.2)), estimate the luminosity ratio L_{He}/L_H between a helium star and a standard hydrogen-abundance star of the same mass.

3. *Opacity for onset of convection*

   An adiabatic gas has pressure and density related through $P \sim \rho^\gamma$, where γ is the adiabatic index.

   a. Use the ideal gas law and hydrostatic equilibrium to show that $|dT/dr|_{ad} = (1 - 1/\gamma)T/H$, where H is the pressure scale height given in Eq. (15.4).

   b. Use the radiative diffusion relations to show that $|dT/dr|_{rad} = \kappa\rho T/(4(\tau + 2/3))$, where τ is the local optical depth.

   c. Combine these to show that the critical opacity for onset of convection is given by $\kappa_c = 4(1 - 1/\gamma)(\tau + 2/3)/(\rho H)$.

4. *Convection within a molecular weight gradient: the Ledoux criterion*

   Consider a medium with a gradient in molecular weight given by $\alpha \equiv d\ln\mu/d\ln r$.

   a. What is true about the molecular weight inside a perturbed fluid element? That is, what is μ_1' in terms of its value μ_0 at the base of the perturbation?

   b. Using this, generalize the analysis illustrated in Figure 17.2 and quantified in Eq. (17.4) to account for a nonzero α.

   c. Show in particular that the condition for onset of convection now becomes $|dT/dr|_{rad} > |dT/dr|_{ad} - \alpha T/r$.

   d. Comment on the physical meaning for how the sign of α determines whether the medium becomes more or less unstable. (This generalization of the Schwarzschild criterion for onset of convection is known as the "Ledoux criterion.")

18 Hydrogen Fusion and the Mass Range of Stars

The timescale analyses in Part I (Chapter 8) show that nuclear fusion of hydrogen into helium provides a long-lasting energy source that we can associate with main-sequence stars in the H–R diagram (Section 6.3). But what are the requirements for such fusion to occur in the stellar core? And how is this to be related to the luminosity versus surface temperature scaling for main-sequence stars in the H–R diagram? In particular, how might this determine the relation between mass and radius? Finally, what does it imply about the lower mass limit for stars to undergo hydrogen fusion (H-fusion)?

18.1 Core Temperature for Hydrogen Fusion

Figure 18.1 illustrates that there are two distinct channels for fusing hydrogen into helium in stellar cores: the direct proton–proton (PP) chain on the left; and the CNO cycle on the right. The latter turns out to be dominant in more-massive stars; their higher core temperatures make it possible for protons to overcome the higher electrical repulsion of the higher charges of CNO nuclei, allowing these to become effective *catalysts* for a net fusion of hydrogen into helium.

But in the Sun and other low-mass stars, the core temperatures are only sufficient for direct PP-chain fusion. The left panel of Figure 18.1 illustrates the most important of the detailed reaction channels, but the overall result is simply

$$4\,_1\mathrm{H}^+ \rightarrow \,_4\mathrm{He}^{+2} + \nu + 2\mathrm{e}^+ + E_\gamma, \tag{18.1}$$

where ν represents a weakly interacting neutrino (which simply escapes the star). The $2\mathrm{e}^+$ represents two positively charged "antielectrons," or *positrons*, which quickly annihilate with ordinary electrons, releasing $\sim 2 \times 2 \times \frac{1}{2} \approx 2\,\mathrm{MeV}$ of energy. The rest of the net $\sim 4 \times 7\,\mathrm{MeV}$ in energy, representing the mass–energy difference between 4 H versus 1 He, is released as high-energy photons (gamma rays) of energy E_γ.

The essential requirement for such PP fusion is that the thermal kinetic energy kT of the protons must overcome the mutual repulsion of their positive charge $+e$, to bring the protons to a close separation at which the strong nuclear (attractive) force is able take over, and bind the protons together. For a given temperature T, the minimum

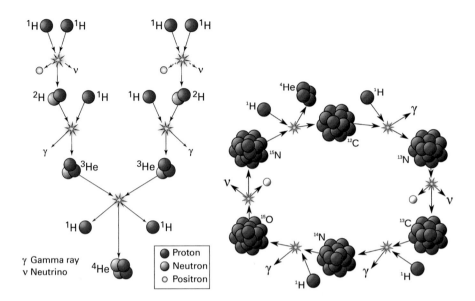

Figure 18.1 The two distinct channels for hydrogen fusion (H-fusion) in stellar cores. For the Sun and other low-mass stars, this occurs by the direct proton–proton (PP) chain (left). For high-mass stars, it occurs via the CNO cycle (right), in which carbon, nitrogen, and oxygen nuclei serve as catalysts for the overall fusion of hydrogen into helium. The higher charge of CNO nuclei requires higher proton energy to overcome the higher electrical repulsion. This makes CNO burning very sensitive to temperature, and so dominant in the hotter cores of higher-mass stars. Credit: Borb.

separation b for two protons colliding head-on comes from setting this thermal kinetic energy equal to the electrostatic repulsion energy,

$$kT = \frac{e^2}{b}. \tag{18.2}$$

In particular, if we were to require that this minimum separation be equal to the size of a helium nucleus, i.e., $b \approx 1 \, \text{fm} = 10^{-15} \, \text{m}$, then from Eq. (18.2) we would infer that the required temperature is quite extreme, $T \approx 1.7 \times 10^{10} \, \text{K}$!

Comparison with the virial scaling (Eq. (15.9)) shows this is more than a 1000 times the characteristic virial temperature for the solar interior, $T_{\text{int}} \approx 13 \, \text{MK}$. As such, the closest distance b between protons in the interior core of the Sun is actually more than 1000 times the size of the helium nucleus, which is thus well outside the scale for operation of the strong nuclear force that keeps the nucleus bound.

The reason that nuclear fusion can nonetheless proceed at such a relatively modest temperature stems again from the uncertainty principle of modern quantum physics. Namely, a proton with thermal energy $m_p v_{\text{th}}^2/2 = kT$ has an associated momentum $p = m_p v_{\text{th}} = \sqrt{2m_p kT}$. Within quantum mechanics, it thus has an associated "fuzziness" in position, characterized by its de Broglie wavelength $\lambda \equiv h/p$, where h is Planck's constant. If $\lambda \gtrsim b$, then there is a good probability that this waviness of protons will allow them to "tunnel" through the electrostatic repulsion barrier between them, and so find themselves within a nuclear distance at which the strong attractive nuclear force can bind them. Setting $b = \lambda = h/(m_p v_{th})$ in Eq. (18.2), we can thus obtain an explicit expression[1] for the proton thermal speed needed for nuclear fusion of hydrogen,

$$ v_{th,\,nuc} = \frac{2e^2}{h} = 690 \text{ km/s.} \tag{18.3} $$

Two remarkable aspects of Eq. (18.3) are: (1) this thermal speed for H-fusion depends *only* on the fundamental physics constants e and h, and (2) its numerical value is very nearly equal to the surface escape speed from the Sun, $v_{esc} = \sqrt{2GM_\odot/R_\odot} = 618$ km/s. Recalling the virial scaling (Eq. (15.9)) that says the thermal energy in the stellar interior is comparable with the gravitational binding energy, this means that, given the solar mass $M_\odot$, the Sun has adjusted to just the radius needed for the gravitational binding to give an interior temperature that is hot enough for H-fusion. For mean molecular weight $\bar{\mu} \approx 0.6 m_p$, the mean thermal speed (Eq. (18.3)) implies a core temperature

$$ T_{nuc} = \frac{\bar{\mu} v_{th,\,nuc}^2}{2k} = 1.2 \frac{m_p e^4}{k h^2} \approx 17 \text{ MK,} \tag{18.4} $$

which now is quite comparable with the interior temperature $T_{int,\,vir} \approx 14$ MK obtained by applying the virial scaling (Eq. (15.9)) to the Sun.

18.2 Main-Sequence Scalings for Radius–Mass and Luminosity–Temperature

If we were naively to apply these same scalings to stars with different masses, then it would suggest all stars along the main sequence should have the same solar ratio of mass to radius, and thus that the radius should increase linearly with mass, $R \sim M$.

In practice, the radius–mass relation for main-sequence stars is somewhat sublinear,

$$ R \sim M^{0.7}. \tag{18.5} $$

This can be understood by considering that the much higher luminosity of more-massive stars, scaling as $L \sim M^3$, means that the core – within which the total fuel available scales just linearly with stellar mass M – must have more vigorous nuclear burning.[2] The higher core temperature to drive such more-vigorous H-fusion then requires by the Virial Theorem that the mass–radius ratio of such stars must be somewhat higher than for lower-mass stars such as the Sun.

---

[1] I am indebted to Professor D. Mullan for pointing out to me this remarkably simple scaling.

[2] Indeed, as already noted, in massive stars the standard, direct proton–proton fusion is augmented by a process called the CNO cycle, in which CNO elements act as a catalyst for H-fusion. Attaching protons to such more-highly charged CNO nuclei requires a higher core temperature to overcome the stronger electrical repulsion, and this indeed obtains in such massive stars.

Combining such a sublinear radius–mass scaling $R \sim M^{0.7}$ with the mass-luminosity scaling $L \sim M^3$ Eq. (17.2) and the Stefan–Boltzmann relation $L \sim T^4 R^2$, we infer that luminosity should be a quite steep function of surface temperature along the main sequence, namely $L \sim T^8$. While observed H–R diagrams (such as plotted for nearby stars in Figure 6.5.) show the main sequence to have some complex curvature structure, a straight line with $\log L \sim 8 \log T$ does give a rough overall fit, thus providing general support for these simple scaling arguments.

18.3   Lower Mass Limit for Hydrogen Fusion: Brown-Dwarf Stars

These nuclear burning scalings can also be used to estimate a minimum stellar mass for H-fusion. Stars with mass below this minimum are known as "brown dwarfs." A key new feature of these stars is that their cores become *electron degenerate*, and so no longer follow the simple virial scalings derived above for stars in which the pressure is set by the ideal gas law. Electron degeneracy occurs when the electron number density n_e becomes comparable to the cube of the electron de Broglie wavenumber $k_e \equiv 2\pi/\lambda_e \equiv 1/\bar{\lambda}_e$,

$$n_e \approx k_e^3 = \frac{1}{\bar{\lambda}_e^3}, \qquad (18.6)$$

with the electron thermal de Broglie (reduced) wavelength,

$$\bar{\lambda}_e = \frac{\hbar}{p_e} = \frac{\hbar}{\sqrt{2m_e kT}}, \qquad (18.7)$$

where $\hbar \equiv h/2\pi$, and the latter equality casts the electron thermal momentum p_e in terms the temperature T and electron mass m_e. Assuming a constant density $\rho = M/(4\pi R^3/3) \approx m_p n_e$, we can combine Eqs. (18.6) and (18.7) with the nuclear temperature in Eq. (18.4) and the virial relation in Eq. (15.9) to obtain a relation for the stellar mass at which a nuclear burning core should become electron degenerate,

$$M_{min, nuc} \approx \frac{0.88}{\pi} \left(\frac{m_p}{m_e}\right)^{3/4} \frac{e^3}{G^{3/2} m_p^2} = \boxed{0.09 M_\odot.} \qquad (18.8)$$

Stars with a mass below this minimum should not be able to ignite H-fusion, because electron degeneracy prevents their cores from contracting to a small enough size to reach the ~ 17 MK temperature (see Eq. (18.4)) required for fusion. In practice, more elaborate computations indicate such brown-dwarf stars have a limiting mass $M_{BD} \lesssim 0.08 M_\odot$, just slightly below the simple estimate given in Eq. (18.8).

Note that, although this minimum mass for H-fusion is limited by electron degeneracy, the actual value is independent of Planck's constant h! Essentially, the role of h in the tunneling effect for H-fusion cancels its role in electron degeneracy.

18.4 Upper Mass Limit for Stars: The Eddington Limit

Now let us consider what sets the *upper* mass limit for observed stars. This is not linked to nuclear burning or degeneracy, but stems from the strong $L \sim M^3$ scaling of luminosity with mass, which, as noted in Section 17.1, follows from the hydrostatic support and radiative diffusion of the stellar envelope.

In addition to its important general role as a carrier of energy, radiation also has an associated momentum. For a photon of energy $E = h\nu$, the associated momentum is set by its energy divided by the speed of light, $p = h\nu/c$. The trapping of radiative energy within a star thus inevitably involves a trapping of its associated momentum, leading to an outward radiative force, or for a given mass, an outward radiative acceleration g_{rad}, that can compete with the star's gravitational acceleration g. For a local radiative energy flux F (energy/time/area), the associated momentum flux (force/area, or pressure) is just F/c. The material acceleration resulting from absorbing this radiation depends on the effective cross-sectional area σ for absorption, divided by the associated material mass m, as characterized by the opacity κ,

$$g_{rad} = \frac{\sigma}{m} \frac{F}{c} = \frac{\kappa F}{c}. \tag{18.9}$$

For a star of luminosity L, the radiative flux at some radial distance r is just $F = L/4\pi r^2$. This gives the radiative acceleration the same inverse-square radial decline as the stellar gravity, $g = GM/r^2$, meaning that it acts as a kind of "anti-gravity."

Sir Arthur Eddington first noted that, even for a minimal case in which the opacity just comes from free-electron scattering, $\kappa = \kappa_e = 0.2(1+X) \approx 0.34 \text{ cm}^2/\text{g}$ (with the numerical value for standard (solar) hydrogen mass fraction $X \approx 0.7$; see Appendix C), there is a limiting luminosity, now known as the *Eddington luminosity*, for which the radiative acceleration $g_{rad} = g$ would completely cancel the stellar gravity,

$$L_{Edd} = \frac{4\pi GMc}{\kappa_e} = 3.8 \times 10^4 L_\odot \frac{M}{M_\odot}. \tag{18.10}$$

Any star with $L > L_{Edd}$ is said to exceed the *Eddington limit*, since even the radiative acceleration from just scattering by free electrons would impart a force that exceeds the stellar gravity, thus implying that the star would no longer be gravitationally bound!

For main-sequence stars that follow the $L \sim M^3$ scaling, setting $L = L_{Edd}$ yields an estimate for an upper mass limit at which the star would reach this Eddington limit,

$$\boxed{M_{max,Edd} \approx 195 \, M_\odot,} \tag{18.11}$$

where $195 \approx \sqrt{3.8 \times 10^4}$. This agrees quite well with modern empirical estimates for the most massive observed stars, which are in the range 150–300 $M_\odot$.

Actually, as stars approach this Eddington limit, the radiation pressure alters the hydrostatic structure of the envelope, causing the mass–luminosity relation to weaken toward a linear scaling, $L \sim M$, and so allowing in principle for stars even with mass $M > M_{max, Edd}$ to remain bound. (See Exercise 6 at the end of this chapter.) In practice, such stars are subject to "photon bubble" instabilities, much as occurs whenever a heavy fluid (in this case the stellar gas) is supported by a lighter one (here the radiation). Very massive stars near this Eddington limit thus tend to be highly variable, often with episodes of large ejection of mass that effectively keeps the stellar mass near or below the $M_{max, Edd} \approx 195 M_{\odot}$ limit.

18.5   Questions and Exercises

Quick Questions

1. Approximately what are the thermal de Broglie wavelengths of electrons and protons in the core of the Sun? Which of these is close to the size of the atomic nucleus?
2. What is the maximum mass M_{max} for a star to have evolved off the main sequence in the 14 Gyr age of the universe?
3. Approximately what is the main-sequence lifetime (Gyr) of a star with mass just above the brown-dwarf limit?

Exercises

1. *Chemically homogeneous evolution*
   Suppose convective instabilities were able to mix hydrogen from the envelope into the core so that the full mass of hydrogen in a star could undergo H-fusion.
   a. How would this affect a star's main-sequence lifetime?
   b. What would be the final composition of such a star? That is, what would be the mass fractions X and Y of hydrogen and helium?
   c. Would such a star evolve into a red giant?
2. *A $100M_{\odot}$ star*
   a. What is the main-sequence luminosity L (in $L_{\odot}$) of a star with mass $M = 100M_{\odot}$?
   b. What is the Eddington luminosity L_{Edd} (in $L_{\odot}$) of such a star?
   c. If the star has a radius $R = 50R_{\odot}$, what is its surface gravity, in m/s^2?
   d. By what factor f does radiation reduce the effective gravity?
   e. Approximately how long (in Myr) would such a star live on the main sequence?
3. *Power-law scaling laws*
   a. Assume a power-law radius–mass scaling $R \sim M^a$ for stars on the main sequence. Show that there is an associated power-law relation $L \sim T^b$ between luminosity L and surface temperature T.
   b. Give a formula for the power index b in terms of the index a.
   c. Compute the values for b for cases with $a = 0.5, 0.7$, and 1.
   d. What value of a would give the $L \sim T^8$ quoted in the text for the main sequence?

4. *Helium stars 3*

   This exercise builds further on the helium-star exercises in Chapters 15 and 17.

   a. For a fully ionized mixture in which the hydrogen, helium, and metal mass fractions are now $X = 0$, $Y = 0.98$, and $Z = 0.02$, compute again the mean molecular weight $\bar{\mu}$ and electron scattering opacity κ_e for such a helium star.

   b. Applying these results in the scaling of Exercise 2 in Chapter 17, again estimate the luminosity ratio L_{He}/L_H between a helium star and a standard hydrogen-abundance star of the same mass.

   c. Now also compute the Eddington luminosity L_{Edd} for such a helium star of mass M.

   d. Taking into account your answers for both parts b and c, now estimate the maximum mass $M_{max, Edd}$ for which such a helium star would reach the Eddington limit $L = L_{Edd}$.

5. *Hydrogen fusion burning*

   a. Using Einstein's famous formula $E = mc^2$, compute the mass reduction rate dM/dt associated with the Sun's radiative luminosity $L_\odot$. Give your answer in both kg/s and $M_\odot$/yr.

   b. Now using the energy release efficiency $\epsilon \approx 0.007$ for hydrogen burning to helium, compute the associated rates of reduction dM_H/dt in the mass of hydrogen in the Sun. Again give your answer in kg/s and $M_\odot$/yr.

   c. Now compute the total decrease in the mass of solar hydrogen (in units of $M_\odot$) over its main-sequence lifetime. What happened to this hydrogen?

   d. The Sun started out with a mass fraction $X = 0.72$ of hydrogen, with the rest mostly helium, with a mass fraction $Y \approx 0.26$. Use these values to compute the average mass fractions X_{tams} and Y_{tams} at the end of the Sun's main-sequence life (known as "tams," for "terminal-age main sequence").

   e. After the main sequence, the Sun will evolve into a red giant with $L \approx 500 L_\odot$, by burning hydrogen in the shell around the hot helium core. How long (in Myr) can it last in this stage before it doubles the amount given in part c for hydrogen consumed by core burning during the full main sequence?

6. *Eddington limit*

   a. Show that the ratio of outward radiation force to gravity is given by the factor $\Gamma \equiv L/L_{Edd}$.

   b. Next show that accounting for this radiation force reduces the stellar effective gravity by a factor $1 - \Gamma$.

   c. Using this to define a reduced effective mass $M_{eff} \equiv M(1 - \Gamma)$, generalize the analysis in Eqs. (17.1) and (17.2) to derive a modified scaling relation for L versus M.

   d. What in particular is the scaling between L and M for $L \lesssim L_{Edd}$?

   e. In this scaling form, can one ever get $L > L_{Edd}$?

19 Post-Main-Sequence Evolution: Low-Mass Stars

As a star ages, more and more of the hydrogen in its core becomes consumed by fusion into helium. Once this core hydrogen is used up, how does the star react and adjust? Without the hydrogen fusion (H-fusion) to supply its luminosity, one might think that perhaps the star would simply shrink, cool, and dim, and so die out, much like a candle when all its wax is used up.

Instead it turns out that stars at this post-main-sequence stage of life actually start to expand at first keeping roughly the same luminosity and so becoming cooler at the surface, but eventually becoming much brighter giant or supergiant stars, shining with a luminosity that can be thousands times that of their core-hydrogen-burning main sequence.

However, as illustrated in Figure 19.1, the detailed evolution and final states of stars depends on the stellar mass, with distinct difference for stars with initial masses below versus above about $8M_\odot$. Figure 19.2 summarizes the corresponding post-main-sequence evolutionary tracks in the H–R diagram for the Sun (top) and for stars with mass up to $10\,M_\odot$ (bottom).

The remainder of this chapter focuses our initial attention on solar-type stars with $M \lesssim 8M_\odot$. The evolution and final states of high-mass stars are discussed in Chapter 20.

19.1   Hydrogen-Shell Burning and Evolution to the Red-Giant Branch

The apparently counterintuitive *post*-main-sequence adjustment of stars can actually be understood through the same basic principles used to understand their initial, *pre*-main-sequence evolution. When the core runs out of hydrogen fuel, the lack of energy generation does indeed cause the core itself to contract. But the result is to make this core even denser and hotter. Then, much as the hot coals at the heart of a wood fire help burn the wood fuel around it much faster, the higher temperature of a contracted stellar core actually makes the overlying *shell* of hydrogen fuel around the core burn even more vigorously!

Such shell burning of hydrogen actually makes the star expand. Unlike during the main sequence – when there is an essential regulation or compatibility between the luminosity generated in the core and the luminosity that the radiative envelope is able to transport to the stellar surface – this shell-burning core is actually overluminous

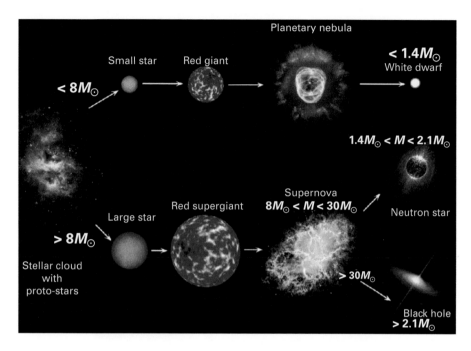

Figure 19.1 Distinct evolution and final states for stars with initial masses above and below $8M_\odot$.

relative to the envelope luminosity that is set by the $L \sim M^3$ scaling law. As such, instead of emitting this core luminosity as surface radiation, the excess energy acts to expand the star, in effect doing work against gravity to reverse the Kelvin–Helmholtz contraction that occurred during the star's pre-main-sequence evolution. Initially, the radiative envelope keeps the luminosity fixed so that, as the star expands, the surface temperature again declines, with the star thus again evolving horizontally on the H–R diagram, this time from left to right.

But as the surface temperature approaches the limiting value $T \approx 3500$–4000 K, the envelope again becomes more and more convective, which thus now allows this full high luminosity of the hydrogen-shell-burning core to be transported to the surface. The star's luminosity thus increases, now with the temperature staying nearly constant at the cool value for the Hayashi limit. In the H–R diagram, the star essentially climbs back up the Hayashi track, eventually reaching the region of the cool, red giants in the upper right of the H–R diagram.

The above describes a general process for all stars, but the specifics depend on the stellar mass. For masses less than the Sun, the main-sequence temperature is already quite close to the cool limit, so evolution can proceed almost directly vertically up the Hayashi track. For masses much greater than the Sun, the luminosity and temperature on the main sequence are both much higher, and so the horizontal evolutionary phase is more sustained. And since the luminosity is already very high, these stars become red supergiants without ever having to reach or climb the Hayashi track.

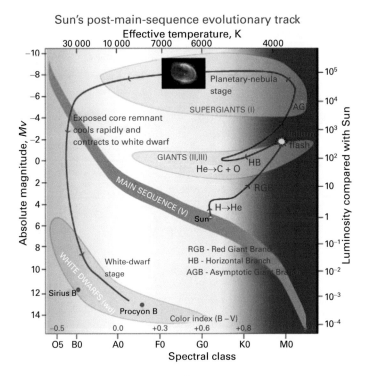

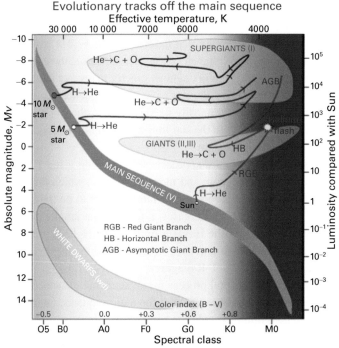

Figure 19.2 Schematic H–R diagrams to show the post-main-sequence evolution for a solar-mass star (top), and for stars with $M = 1, 5$, and $10 M_\odot$ (bottom). Credit: CSIRO Radio Astronomy Image Archive.

19.2 Helium Flash to Horizontal Branch Core Burning

This hydrogen-shell burning also has the effect of increasing further the temperature of the stellar core, and eventually this reaches a level where the fusion of the helium itself becomes possible, through what is known as the "triple-α process,"[1]

$$3 \, _4\text{He}^{+2} \leftrightharpoons \, _8\text{Be}^{+4} +_4 \text{He}^{+2} \rightarrow \, _{12}\text{C}^{+6}. \tag{19.1}$$

The direct fusion of two $_4\text{He}^{+2}$ nuclei initially makes an unstable nucleus of berylium ($_8\text{Be}^{+4}$), which usually just decays back into helium. But if the density and temperature are sufficiently high, then during the brief lifetime of the unstable berylium nucleus, another helium can fuse with it to make a very stable carbon nucleus $_{12}\text{C}^{+6}$. Since the final step of fusing $_4\text{He}^{+2}$ and $_8\text{Be}^{+4}$ involves overcoming an electrostatic repulsion that is $2 \times 4 = 8$ times higher than for proton–proton (PP) fusion of hydrogen, helium fusion (He-fusion) requires a much higher core temperature, $T_{\text{He}} \approx 8 \times 15\,\text{MK} \approx 120\,\text{MK}$.

In stars with more than a few solar masses, this ignition of the helium in the core occurs gradually, since the higher core temperature from the addition of helium burning (He-burning) increases the gas pressure, making the core tend to expand in a way that regulates the burning rate.

In contrast, for the Sun and other stars with masses $M < 2M_\odot$, the number density of electrons n_e in the helium core is so high[2] that their core becomes *electron degenerate*. As discussed in Section 18.3 for the brown-dwarf stars that define the lower mass limit for hydrogen burning (H-burning), electron degeneracy occurs when the mean distance between electrons $\sim n_e^{-1/3}$ becomes comparable to the de Broglie wavelength $\bar{\lambda}_e = \bar{h}/p_e$. The properties of such degeneracy are discussed further in Section 19.4 on the degenerate white-dwarf end-states of solar-type stars.

But in the present context, a key point is that the response to any heat addition is quite different than for an ideal gas. By the Virial Theorem for a gravitationally bound ideal gas, the added heating from any increase in nuclear burning leads to an expansion that cools the gas, thus reducing the burning and so keeping it stable. In contrast, for a degenerate gas, the expansion from adding heat actually makes the temperature increase even further. Thus, once the evolutionary increase in core temperature reaches a level that ignites fusion of helium into carbon, the degenerate nature of the gas leads to a *helium flash*, in which a substantial fraction of the core of helium is fused into carbon over a very short timescale.

As illustrated in the top panel of Figure 19.2, this flash marks the "tip" of the Red Giant Branch (RGB) in the H–R diagram. Somewhat surprisingly, the sudden

---

[1] Because helium nuclei are sometimes referred as "α-particles." In this formula, the left subscript denotes the atomic mass (number of proton and neutrons), while the right superscript denotes the nuclear charge (number of protons).

[2] Recall that, on the main sequence, the radii of stars is (very) roughly proportional to their mass, $R \sim M$. But since density scales as $\rho \sim M/R^3$, the *density* of low-mass stars tends generally to be higher than in high-mass stars, roughly scaling as $\rho \sim 1/M^2$. This overall scaling of average stellar density also applies to the relative densities of stellar cores, and so helps to explain why the cores of low-mass stars tend to become electron degenerate, while those of higher-mass stars do not.

addition of energy is largely absorbed by the expansion of the core and the overlying stellar envelope. Since the expanded core is no longer very degenerate, the star thus simply settles down to a more quiescent, stable phase of He-burning. The expanded core also means the shell burning of hydogen actually declines, causing the luminosity to decrease from the tip of the Red Giant Branch, where the helium flash occurs, to a somewhat hotter, dimmer region known as the Horizontal Branch in the H–R diagram.

This Horizontal Branch (HB) can be loosely thought of as the He-burning analog of the H-burning main sequence, but a key difference is that it lasts a much shorter time, typically only 10 to 100 *million years*, much less than the many billion years for a solar mass star on the main sequence. This is partly because the luminosity for HB stars is so much higher than for a similar mass on the main sequence, implying a much higher burn rate of fuel. But another factor is that the energy yield per-unit-mass, ϵ, for He-fusion to carbon is about a tenth of that for H-fusion to helium, namely about $\epsilon_{He} \approx 0.06$ percent versus the $\epsilon_H \approx 0.7$ percent for H-burning (see Figure 20.1). With the lower energy produced, and the higher rate of energy lost in luminosity, the lifetime is accordingly shorter.

19.3   Asymptotic Giant Branch to Planetary Nebula to White Dwarf

Once the core runs out of helium, He-burning also shifts to an inner shell around the core, which itself is still surrounded by a outer shell of more vigorous H-burning. This again tends to increase the core luminosity, but now, because the star is cool and thus mostly convective, this energy is mostly transported to the surface with only a modest further expansion of the stellar radius. This causes the star to again climb in luminosity along what is called the Asymptotic Giant Branch (AGB), which parallels the Hayashi track at just a somewhat hotter surface temperature.

In the Sun and stars of somewhat higher mass, up to $M \lesssim 8M_\odot$, there can be further ignition of the carbon fusing with helium to form oxygen. But further synthesis up the periodic table requires overcoming the greater electrical repulsion of more highly charged nuclei. This, in turn, requires a temperature higher than occurs in the cores of lower-mass stars, for which the onset of electron degeneracy prevents contraction to a denser, hotter core. Further core burning thus ceases, leaving the core as an inert, degenerate ball of carbon and oxygen, with final mass on order of $1\,M_\odot$, with the remaining mass contained in the surrounding envelope of mostly hydrogen.

But such AGB stars tend also to be pulsationally unstable and, because of the very low surface gravity, such pulsations can over time actually eject the entire stellar envelope. This forms a circumstellar *nebula* that is heated and ionized by the very hot remnant core. As seen in the left panel of Figure 20.4, the resulting circular nebular emission glow somewhat resembles the visible disk of a planet, so these are called "planetary nebulae," though they really have nothing much to do with actual planets. After a few thousand years, the planetary nebula dissipates, leaving behind just the degenerate remnant core, a white-dwarf star.

19.4 White-Dwarf Stars

The electron-degenerate nature of white-dwarf stars endows them with some rather peculiar, even extreme, properties. As noted, they typically consist of roughly a solar mass of carbon and oxygen, but have a radius comparable to that of the Earth, $R_e \approx R_\odot/100$. This small radius makes them very dense, with $\rho_{wd} \approx 10^6 \bar{\rho}_\odot \approx 10^6$ g/cm^3, i.e., about a million times (!) the density of water, and so a million times the density of normal main-sequence stars such as the Sun. It also gives them very strong surface gravity, with $g_{wd} \approx 10^4 g_\odot \approx 10^6$ m/s^2, or about 100 000 times Earth's gravity!

As noted in Section 18.3 for the brown-dwarf stars that define the lower mass limit for H-burning, a gas becomes electron degenerate when the electron number density n_e becomes so high that the mean distance between electrons becomes comparable to their reduced de Broglie wavelength,

$$n_e^{-1/3} \approx \bar{\lambda} \equiv \frac{\hbar}{p_e} = \frac{\hbar}{m_e v_e}, \tag{19.2}$$

where the electron thermal momentum p_e equals the product of its mass m_e and thermal speed v_e, and $\hbar \equiv h/2\pi$ is the reduced Planck constant. The associated electron pressure is

$$P_e = n_e v_e p_e = n_e^{5/3} \frac{\hbar^2}{m_e} = \left(\frac{\rho Z}{A m_p}\right)^{5/3} \frac{\hbar^2}{m_e}, \tag{19.3}$$

where the last equality uses the relation between electron density and mass density, $\rho = n_e A m_p/Z$, with Z and $A m_p$ the average nuclear charge and atomic mass. For example, for a carbon white dwarf, the atomic number $Z = 6$ gives the number of free (ionized) electrons needed to balance the $+Z$ charge of the carbon nucleus, while the atomic weight $A m_p = 12 m_p$ gives the associated mass from the carbon atoms. The hydrostatic equilibrium (cf. Eq. (15.1)) for pressure-gradient support against gravity then requires, for a white-dwarf star with mass M_{wd} and radius R_{wd},

$$\frac{P_e}{R_{wd}} \approx \rho \frac{G M_{wd}}{R_{wd}^2}. \tag{19.4}$$

Using the density scaling $\rho \sim M_{wd}/R_{wd}^3$, we can combine Eqs. (19.3) and (19.4) to solve for a relation between the white-dwarf radius and its mass,

$$R_{wd} \approx \frac{3.6}{G M_{wd}^{1/3}} \frac{\hbar^2}{m_e} \left(\frac{Z}{A m_p}\right)^{5/3} \approx \boxed{0.01 R_\odot \left(\frac{M_\odot}{M_{wd}}\right)^{1/3}}, \tag{19.5}$$

where the approximate evaluation uses the fact that for both carbon and oxygen the ratio $Z/A = 1/2$, and the factor 3.6 comes from a detailed computation not covered here. For a typical mass of order the solar mass, we see that a white dwarf is very compact, comparable to the radius of the Earth, $R_e \approx 0.01 R_\odot$. But note that this radius actually *decreases* with increasing mass.

19.5 Chandrasekhar Limit for White-Dwarf Mass: $M < 1.4 M_\odot$

This fact that white-dwarf radii decrease with higher mass means that, to provide the higher pressure to support the stronger gravity, the electron speed v_e must strongly increase with mass. Indeed, at some point this speed approaches the speed of light, $v_e \approx c$, implying that the associated electron pressure now takes the scaling (cf. Eq. (19.3)),

$$P_e = n_e c p_e = n_e^{4/3} \hbar c = \left(\frac{\rho Z}{A m_p} \right)^{4/3} \hbar c. \tag{19.6}$$

Applying this in the hydrostatic relation (19.4), we now find that the radius R cancels! Instead we can solve for an upper limit for a white-dwarf's mass,

$$M_{wd} \leq M_{ch} = \sqrt{3\pi} \left(\frac{\hbar c}{G} \right)^{3/2} \left(\frac{Z}{A m_p} \right)^2 \approx \boxed{1.4 M_\odot,} \tag{19.7}$$

where the subscript refers to "Chandrasekhar," the astrophysicist who first derived this mass limit, and the proportionality factor $\sqrt{3\pi}$ comes from a detailed calculation beyond the scope of the discussion here.

As discussed in later sections (see, for example, Section 31.1), when accretion of matter from a binary companion puts a white dwarf over this limit, it triggers an enormous "white-dwarf supernova" explosion, with a large, relatively well-defined peak luminosity, $L \approx 10^{10} L_\odot$. This provides a very bright standard candle that can be used to determine distances as far as 1 Gpc, giving a key way to calibrate the expansion rate of the universe.

But in the present context, this limit means that sufficiently massive stars with cores above this mass cannot end their lives as a white dwarf. Instead, they end as violent "core-collapse supernovae," leaving behind an even more-compact final remnant, either a neutron star or black hole, as we discuss in the next chapter.

19.6 Questions and Exercises

Quick Questions

1. Note that pressure can be written as $P = Nvp$ for speed v and momentum $p = mv$ in a medium with N atoms of mass m. Show that this recovers the ideal gas form for pressure by identifying the thermal speed scaling with temperature, $kT/2 = mv_{th}^2/2$.

2. Derive a general formula for the escape speed v_{esc} for a white dwarf of mass M. What is the value (in km/s) for $M = M_\odot$ and how does this compare with the Sun's escape speed?

3. What is the maximum escape speed (in km/s) of a white dwarf?

4. Derive a general formula for $\log g$ of a white-dwarf star of mass M, and compare this with $\log g_\odot$.

Exercises

1. *Planetary nebula emission and expansion*

   Suppose we observe the Hα line from a spherical planetary nebula, and find it has a maximum wavelength of 656.32 nm and minimum wavelength of 656.24.

   a. Estimate how fast the nebula is expanding. Give your answer in both km/s and au/year.

   b. Now suppose that over a time of 10 years the nebula's angular diameter is observed to expand from 10 arcsec to 10.1 arcsec. What is the distance to the nebula (in pc)?

   c. What is actual physical diameter of the nebula (in au)?

2. *White-dwarf cooling time*

   a. How many carbon nuclei are there in a pure-carbon white dwarf of mass 1 $M_\odot$?

   b. If the carbon is full ionized, how many electrons are there?

   c. Each particle in a gas (even a degenerate gas) of temperature T has a thermal energy $(3/2)kT$. If this fully ionized carbon white dwarf has a constant temperature of 10^7 K through nearly all of its interior mass, what is its total thermal energy (in J)?

   d. Assuming it has a radius of 0.01 $R_\odot$ and radiates like a 30 000 K blackbody, compute its luminosity, L_{wd} (in $L_\odot$).

   e. Now use these results to estimate this white dwarf's cooling lifetime, t_{cool} (in yr).

3. *White-dwarf supernova*

   Consider a pure-carbon white dwarf at the Chandrasekhar mass limit that collapses suddenly from a radius $R \approx R_e$, igniting a sudden conversion of the carbon to iron.

   a. Considering the relative binding energy per nucleon of iron versus carbon (e.g., from Figure 20.1), estimate the associated nuclear energy release E_n (in J) if the entire stellar mass is converted to iron.

   b. Assuming the initial white dwarf has a uniform density, compute its gravitational binding energy E_g, and compare this with the nuclear energy release E_n from part a.

   c. Assuming just 1 percent of this nuclear energy is radiated as light within two weeks of the collapse, estimate the average luminosity L_{avg}, in units of $L_\odot$.

   d. If the *peak* luminosity during this time is twice this average (i.e. $L_{peak} = 2L_{avg}$), what is the absolute magnitude, M_{peak}, at peak brightness?

   e. Using a telescope with a detection limit of apparent magnitude $m = +20$, what is maximum distance d (in Mpc) that the peak brightness of such an explosion can be detected?

4. *Helium stars 4*

   This exercise builds further on the helium-star exercises in Chapters 15, 17, and 18.

   a. For a fully ionized mixture in which the hydrogen, helium, and metal mass fractions are now $X = 0$, $Y = 0.98$, and $Z = 0.02$, compute again the mean molecular weight $\bar{\mu}$ for such a helium star.

b. Use this to generalize Eq. (15.9) to give the scaling of the interior temperature T_{int} of a helium star as a function of its mass M and radius R.

c. As discussed following Eq. (19.1), assume that fusion of He $\rightarrow$ C requires a temperature that is a factor $4 \times 2 = 8$ higher than the $\sim 15\,\text{MK}$ for H-fusion. Now set this $T_{He-fusion} = T_{int}$ to derive a radius–mass scaling $R(M)$ for such stars on the "helium main sequence."

d. Again use the scaling in Chapter 17 to derive helium-modified mass–luminosity scaling, $L(M)$.

e. Combine the results for c and d to estimate for such helium-main-sequence values for the constants A and b in the scaling relation $L_{HeMS}/L_\odot \approx A(T/T_\odot)^b$.

5. *Algol paradox and binary mass exchange*
   The system Algol includes a short period ($P = 2.87\,\text{days}$) binary consisting of an early-type main-sequence star (Aa1, B8V, $M = 3.2M_\odot$) and a later-type subgiant (Aa2, K0IV, $M = 0.7M_\odot$). The orbit is nearly circular.

   a. Explain why this represents a "paradox" in the context of single-star evolution.

   b. What is the binary separation $a = a_1 + a_2$ (in au)?

   c. What are the distances of each component a_1, a_2 from the center of mass?

   d. Relative to this center of mass, what is the location of the Roche point between the stars? (See Exercise 2 of Chapter 7.)

   e. Suppose the mass ratio between the stars was once reversed; what could happen when the initially more-massive star evolved from its main sequence, to a radius that reaches this Roche point?

   f. Discuss how this can resolve the Algol paradox.

20 Post-Main-Sequence Evolution: High-Mass Stars

20.1 Multiple Shell Burning and Horizontal Loops in the H–R Diagram

The post-main-sequence evolution of stars with higher initial mass, $M > 8M_\odot$ has some distinct differences from that outlined in previous chapters for solar- and intermediate-mass stars. Upon exhaustion of hydrogen fuel (H-fuel) at the end of the main sequence, such stars again expand in radius because of the overluminosity of H-shell burning. But the high luminosity and high surface temperature on the main sequence means that their stellar envelopes remain radiative even as they expand, never reaching the cool temperatures that force a climb up the Hayashi track. Instead, their evolution tends to keep near the constant luminosity set by $L \sim M^3$ scaling for the star's mass, so evolving horizontally to the right on the Hertzsprung–Russell diagram.

Because stellar radii scale nearly linearly with mass $R \sim M$, the mean stellar density $\rho \sim M/R^3 \sim M^{-2}$ tends to decline with increasing mass. Thus, even after the core contraction that occurs toward the end of nuclear burning, the core density of high-mass stars never becomes high enough to become electron degenerate. Moreover, the higher mass means a stronger gravitational confinement that gives higher central temperature and pressure. This now makes it possible to overcome the increasingly strong electrical repulsion of more highly charged, higher elements, allowing nucleosynthesis to proceed up the period table all the way to iron, which is the most stable nucleus.

However, each such higher level of nucleosynthesis yields proportionally less and less energy. This can be seen from the plot in Figure 20.1 of the binding energy per nucleon versus the number of nucleons in a nucleus. The jump from hydrogen to helium yields 7.1 MeV, which relative to a nucleon mass of 931 MeV represents a percentage energy release of about 0.7 percent, as noted in Section 19.2. But from helium to carbon the release is just $7.7 - 7.1 = 0.6$ MeV, representing an energy efficiency of just 0.06 percent. As the curve flattens out, the fractional energy release become even less until, for elements beyond iron, further fusion would require the *addition* of energy.

For such massive stars, the final stages of post-main-sequence evolution are characterized by an increasingly massive iron core that can no longer produce any energy by further fusion. But fusion still occurs in a surrounding series of shells, somewhat like the onion-skin structure shown in Figure 20.2, with higher elements fusing in

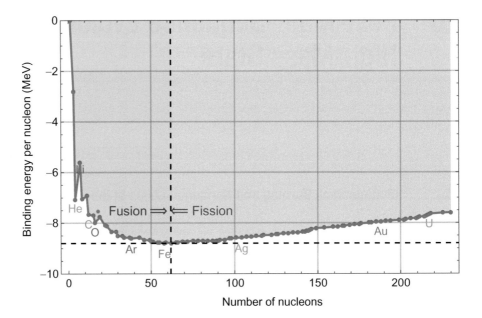

Figure 20.1 Binding energy per nucleon plotted versus the number of nucleons in the atomic nucleus. The *fusion* of light elements moves nuclei to the right, releasing the energy of nuclear burning in the very hot dense cores of stars, but only up to formation of the most stable nucleus, just beyond iron (Fe), with atomic number $A = 56$. Heavier elements are produced in the sudden core collapse of massive-star supernovae, and by merger of binary neutron stars (see Figure 20.7). The *fission* of such heavy elements leads to lower-mass nuclei toward the left. The energy released is what powers nuclear fission reactors here on Earth. Adapted from original graphic by Keith Gibbs at www.schoolphysics.co.uk/.

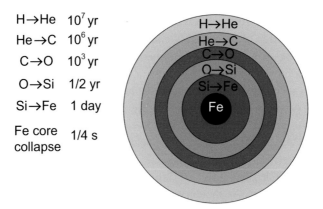

H→He	10^7 yr
He→C	10^6 yr
C→O	10^3 yr
O→Si	1/2 yr
Si→Fe	1 day
Fe core collapse	1/4 s

Figure 20.2 The "onion-skin" layering of the core of a $\sim 20\ M_\odot$ star just before supernovae core collapse, illustrating the various stages of nuclear burning in shells around the inert iron core. The table on the left shows the decreasing duration for each higher stage of burning.

the innermost, hottest shells, and outer shells fusing lower elements, extending to an outermost shell of hydrogen burning.

20.2   Core-Collapse Supernovae

With the build-up of iron in the core, there is an increasingly strong gravity, and without any further fusion-generated energy, the core contracts, ultimately reaching such high energy that electrons enter nuclei and join with protons to create free neutrons. Since neutron masses are much greater than that of electrons, this makes the gas much less degenerate and relativistic, while also removing energy and so reducing pressure support, thus exacerbating the contraction and producing evermore neutrons. The resulting catastrophic "core collapse" continues until the density becomes so high that the neutrons themselves become degenerate, at a core size of order a few tens of kilometers. The "stiffness" of this neutron-degenerate core leads to a "rebound" in the collapse, with gravitational release from the core contraction now powering an explosion that blows away the entire outer regions of the star,[1] with the stellar ejecta reaching speeds of about 10 percent the speed of light!

This ejecta contains iron and other heavy elements, including some beyond iron that are fused in less than a second (1 s) of the explosion by the enormous energy and temperatures. While elements up to oxygen can also be synthesized in low-mass stars, the heavier elements up to iron are thought to have originated in such core-collapse supernova explosions.[2] For a few weeks, the luminosity of such a supernova can equal or exceed that of a whole galaxy, up to $\sim 10^{12} L_\odot$!

Though the dividing line is not exact, it is thought (see Figure 19.1) that all stars with initial masses $M > 8M_\odot$ will end their lives with such a core-collapse supernova, instead of following the track, AGB $\rightarrow$ PN $\rightarrow$ WD (white dwarf), for stars with initial mass $M < 8M_\odot$. Stars with initial masses $8M_\odot < M \lesssim 30M_\odot$ are thought to leave behind a *neutron-star* remnant, as discussed in Section 20.3. But such neutron-star remnants have their own upper mass limit of $M_{ns} \lesssim 2.1M_\odot$, beyond which the gravity becomes so strong that not even the combination of nuclear forces and degenerate pressure from neutrons can prevent a further collapse, this time forming a *black hole*. This is thought to be the final core remnant for the most massive stars, those with initial mass $M \gtrsim 30M_\odot$.

---

[1] Actually, in many simulations, this rebound stalls, and the material recollapses to directly form a black hole, without much of an external explosion; there is now some evidence from large-scale surveys that some uncertain fraction of massive stars may indeed just end their lives by such collapse into a black hole without a visible superova explosion.

[2] Neutron-rich elements beyond iron are now believed to be primarily produced in "kilonovae" that arise from merger of binary neutron stars; see Section 20.6 and Figure 20.7.

20.3   Neutron Stars

Neutron stars are even more bizarrely extreme than white dwarfs. With a mass typically about twice that of the Sun, they have a radius comparable to a small city, $R_{\rm ns} \approx 10\,{\rm km}$, about a factor 600 smaller than even a white dwarf, implying a density that is about 10^8 times higher, and a surface gravity more than 10^5 times higher.

Their support against this very strong gravity comes from both nuclear forces and *neutron* degeneracy pressure, a combination that makes computation of their internal structure very challenging and a topic of much current research. But their overall properties can be well estimated by a procedure for treating neutron degeneracy in a way that is quite analogous to that used in Sections 19.4 and 19.5 for white dwarfs supported by electron degeneracy, just substituting now the electron mass with the *neutron* mass, $m_{\rm e} \rightarrow m_{\rm n} \approx m_{\rm p}$, and setting $Z/A = 1$. The radius–mass relation thus now becomes (cf. Eq. (19.5)),

$$R_{\rm ns} = 2^{5/3} \frac{m_{\rm e}}{m_{\rm p}} R_{\rm wd} = \frac{3.6}{G M_{\rm ns}^{1/3}} \frac{\hbar^2}{m_{\rm p}^{8/3}} \approx \boxed{10\,{\rm km} \left(\frac{M_\odot}{M_{\rm ns}} \right)^{1/3}.} \tag{20.1}$$

Note again that, as in the case of an electron-degenerate white dwarf, this neutron-star radius also *decreases* with increasing mass.

For analogous reasons that lead to the upper mass limit for white dwarfs, for sufficiently high mass the neutrons become relativistic, leading now to an upper mass limit for neutron stars (cf. Eq. (19.7)) that scales as

$$M_{\rm ns} \le M_{\rm lim} = 1.1 \left(\frac{\hbar c}{G} \right)^{3/2} \left(\frac{1}{m_{\rm p}} \right)^2 \approx \boxed{2.1 M_\odot,} \tag{20.2}$$

where again the factor 1.1 comes from detailed calculations not covered here; apart from this and the factor $(A/Z)^2 = 4$, this "Tolman–Oppenheimer–Volkoff" (TOV) limit is the same form as the Chandrasekhar limit for white dwarfs in Eq. (19.7). The exact value of this limiting mass is still a matter of current research, with the value quoted here just somewhat below the value $2.2 M_\odot$ inferred from gravitational waves detected from merger of two neutron stars; see Section 20.6. Neutron stars above this limiting mass will again collapse, this time forming a *black hole*.

20.4   Black Holes

Black holes are objects for which the gravity is so strong that not even light itself can escape. A proper treatment requires General Relativity, Einstein's radical theory of gravity that supplants Newton's theory of universal gravitation, and extends it to the limit of very strong gravity. But we can nonetheless use Newton's theory to derive some basic scalings. In particular, for a given mass M, a characteristic radius for which the Newtonian escape speed is equal to speed of light, $v_{esc} = c$, is just

$$R_{bh} = \frac{2GM}{c^2} \approx 3 \,\text{km} \, \frac{M}{M_\odot},$$ (20.3)

which is commonly known as the "Schwarzschild radius."

Because the speed of light is the highest speed possible, any object within this Schwarzschild radius of a given mass M can *never escape* the gravitational binding from that mass. In terms of Einstein's General Theory of Relativity, mass acts to bend space and time, much the way a bowling ball bends the surface of a trampoline. And much as a sufficiently dense, heavy ball could rip a hole in the trampoline, for objects with mass concentrated within a radius R_{bh}, the bending becomes so extreme that it effectively punctures a hole in spacetime. Since not even light can ever escape from this hole, it is completely black, absorbing any light or matter that falls in, but never emitting any light from the hole itself. This the origin of the term "black hole."

Stellar-mass black holes with $M \gtrsim 2.1 M_\odot$ form from the deaths of massive stars. If left over from a single star, they are hard or even impossible to detect, since by definition they do not emit light.

However, in a binary system, the presence of a black hole can be indirectly inferred by observing the orbital motion (visually or spectroscopically via the Doppler effect) of the luminous companion star.

Moreover, when that companion star becomes a giant, it can, if it is close enough, transfer mass onto the black hole, as illustrated in Figure 20.3. Rather than falling directly into the hole, the conservation of the angular momentum from the stellar orbit requires that the matter must first feed an orbiting *accretion disk*. By the Virial Theorem, half the gravitational energy goes into kinetic energy of orbit, but the other half is dissipated to heat the disk, which, by the blackbody law, then emits it as radiation.

The luminosity of such black-hole accretion disks can be very large. For a black hole of mass M_{bh} accreting mass at a rate $\dot{M}_a$ to a radius R_a that is near the Schwarzschild radius R_{bh}, the luminosity generated is

$$L_{disk} = \frac{GM_{bh}\dot{M}_a}{2R_a} = \frac{R_{bh}}{4R_a}\dot{M}_a c^2 \equiv \epsilon \dot{M}_a c^2.$$ (20.4)

The latter two equalities define the efficiency $\epsilon \equiv R_{bh}/4R_a$ for converting the rest-mass energy of the accreted matter into luminosity. For accretion radii approaching the Schwarzschild radius, $R_a \approx R_{bh}$, this efficiency can be as high as $\epsilon \approx 0.25$, implying 25 percent of the accreted matter-energy is converted into radiation. By comparison, for H-fusion of a main-sequence star the overall conversion efficiency is about 0.07 percent, representing the ~ 10 percent core mass that is sufficiently hot for H-fusion at a specific efficiency, $\epsilon_H = 0.007 = 0.7$ percent.

In the inner disk, the associated blackbody temperature can reach 10^7 K or more; and at the very inner disk edge, dissipation of the orbital energy can heat material to even more extreme temperatures, up to $\sim 10^{10}$ K. By studying the resulting

Figure 20.3 Artist depiction of mass transfer onto an accretion disk around a black hole in Cygnus X-1, wherein the power generated by falling through the strong gravitational potential heats the disk to X-ray-emitting temperatures, while also driving a bipolar jet of high-energy particles. Credit NASA/CXO.

high-energy radiation, we can infer the presence and basic properties (mass, even rotation rate) of black holes in such binary systems, even though we cannot see the black hole itself.

Figure 20.3 shows an artist depiction of the mass transfer accretion in the high-mass X-ray binary Cygnus X-1, thought to be the clearest example of a stellar-remnant black hole, estimated in this case to have a mass $M_{bh} > 10 M_\odot$ that is well above the $M_{lim} \approx 2.1\ M_\odot$ upper limit for a neutron star (cf. Eq. (20.2)).

20.5   Observations of Stellar Remnants

It is possible to observe directly all three types of stellar remnants.

1. *Planetary nebula and white-dwarf stars*

   Stars with initial mass $M < 8 M_\odot$ evolve to an asymptotic giant branch (AGB) star that ejects the outer stellar envelope to form a planetary nebula (PN) with the hot stellar core with mass below the Chandrasekhar mass, $M_{wd} < M_{ch} = 1.4 M_\odot$. Once the nebula dissipates, this leaves behind a white dwarf (WD). White-dwarf stars are very hot, but with such a small radius that their luminosity is very low, placing them on the lower left of the H–R diagram.

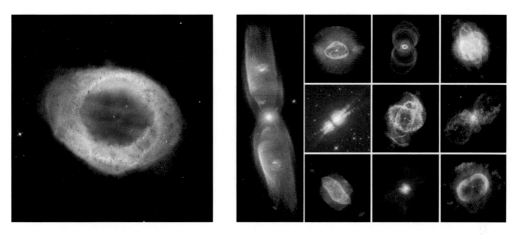

Figure 20.4 Left: M57, known as the Ring Nebula, provides a vivid example of a spherically symmetric planetary nebula. The central hot star is the remnant of the stellar core, and after the nebula dissipates, it will be left as a white-dwarf star. Credit: NASA, ESA and the Hubble Heritage Collaboration. Right: A gallery of planetary nebulae, showing the remarkable variety of shapes that probably stem from interaction of the stellar ejecta with a binary companion, or perhaps even with the original star's planetary system. Credit: NASA/HST.

The excitation and ionization of the gas in the surrounding planetary nebula makes it shine with an emission line spectrum, with the wavelength-specific emission of various ion species giving it a range of vivid colors or hues. Figure 20.4 shows that these planetary nebulae can thus be visually quite striking, with spherical emission nebula from single stars (left), or very complex geometric forms (right) for stars in binary systems.

2. *Neutron stars and pulsars*

A star with initial masses in the range $8M_\odot < M \lesssim 30M_\odot$ ends its life as a core-collapse supernova that leaves behind a neutron star with mass $1.4M_\odot < M_{ns} < 2.1M_\odot$. The conservation of angular momentum during the collapse to such a small size (~ 10 km) makes them rotate very rapidly, often many times a second! This also generates a strong magnetic field, and when the polar axis of this field points toward Earth, it emits a strong *pulse* of beamed radiation in the radio to optical to even X-rays. This is observed as a *pulsar*.

One of the best-known examples is the Crab pulsar, which lies at the center of the Crab Nebula, the remnant from a core-collapse supernova that was observed by Chinese astronomers in 1054 AD. Figure 20.5 shows images of this Crab Nebula in the optical region (left) and in a composite of images (right) in the optical (red) and X-ray (blue) wavebands.

3. *Black holes and X-ray binary systems*

Finally, stars with initial masses $M \gtrsim 30M_\odot$ end their lives with a core collapse supernova that now leaves behind a black hole with mass $M_{bh} > 2.1M_\odot$. As noted, in single stars, these are difficult or impossible to observe, because they emit no light; but in binary systems, accretion from the other star can power a

Figure 20.5 Left: Optical image of the Crab Nebula, showing the remnant from a core-collapse supernova whose explosion was observed by Chinese astronomers in 1054 AD. Credits: NASA, ESA, J. Hester and A. Loll (Arizona State University). Right: A composite zoomed-in image of the central region Crab Nebula, showing Chandra X-rays (blue and white), Hubble visible (purple), and Spitzer infrared (pink) data. The bright star at the nebular center is the Crab pulsar, a rapidly rotating neutron star that was left over from the supernova explosion. Credit: NASA/CXO/HST/Spitzer.

bright accretion disk around the black hole that radiates in high-energy bands like X-rays and even gamma rays. Figure 20.3 shows an artist depiction of accretion onto a black hole in the high-mass X-ray binary Cygnus X1. As noted in Section 13.2, the Event Horizon Telescope has recently imaged the millimeter-wave emission from a supermassive black hole at the center of the galaxy M87; see also Section 26.4.

20.6   Gravitational Waves from Merging Black Holes or Neutron Stars

Einstein's publication in 1915 of his General Theory of Relativity cast gravity as the bending of spacetime. It also directly implied that an *accelerating* mass would generate a wave in the changing gravity, propagating at the speed of light. Because gravity is so weak, Einstein himself never thought such *gravitational waves* could be detectable.

Indeed, even for the strongest imagined sources – the merger of two stellar-mass black holes – the expected signal after traveling a characteristic distance to Earth is estimated to be very weak, alternatively stretching and compressing distances by a tiny relative fraction of just 10^{-21}! Thus, for example, over a length of 1 km, this would require measurement of length changes on a scale of 10^{-18} m, or about 1/1000 the size of a proton!

Nonetheless, through a remarkable combination of ingenuity and heroic scientific ambition, designs were developed over several decades that led to the construction in 2002 of an instrument called LIGO (Laser Interferometer Gravitational-wave Observatory), designed to detect waves from mergers of compact objects such as black holes and neutron stars. Following extended further development over more than a decade to improve the sensitivity and reduce noise, a version called Advanced LIGO finally detected gravitational waves from the merger of two black holes. After traveling over a billion light years, these waves arrived at Earth and the LIGO detectors on September 17, 2015, one century after Einstein's publication of the General Relativity theory that predicted their existence.

As illustrated in Figure 20.6, LIGO was able to detect[3] these tiny deflections by analyzing the interference pattern between two laser signals that reflect from mirrors at the end of two perpendicular, 4 km-long arms. Comparing variations from duplicate detectors in both Louisiana and Washington states helped discriminate against false signals from local disturbances and noise sources. As the black holes spiraled ever closer toward merger, the waves steadily increased in frequency, which when translated into sound gave a characteristic "chirp." (See lower panel of Figure 20.6.) When analyzed in comparison to computer simulations of such mergers, the chirp pattern indicated the black holes in this first-detected merger were quite massive, about $29 M_\odot$ and $36 M_\odot$, several times higher than the most-massive black holes inferred in the high-mass X-ray binaries discussed in Section 20.5. The final, merged black hole was inferred to be about $62 M_\odot$, with the extra $3 M_\odot$ converted to energy in the emitted gravitational wave, which for a brief few milliseconds of the merger represented some 50 times the luminosity of all the stars in the observable universe!

Just two years after this historic discovery, the 2017 Nobel Prize in Physics was awarded to three leaders of the team that developed LIGO. This followed on from the 1992 Noble Prize, awarded to two astronomers who, in 1982, discovered a binary system that provided strong indirect evidence for gravitational waves. The measured changes in the period of this pulsar orbiting a neutron star closely followed the predicted changes from orbital decay associated with loss of orbital energy through emission of gravitational waves.

In August 2017, LIGO then detected the first *merger* of such neutron stars in close binary orbit. Unlike the merger of black holes, which owing to their restriction against any light emission had no detected electromagnetic (EM) signatures, this neutron-star merger was also detected in EM spectral bands ranging from gamma rays and X-rays, to ultraviolet and optical light, to infrared and even radio waves. In particular, just 1.7 s after the recorded gravitational wave, a 2 s burst of gamma rays was observed by the Fermi and INTEGRAL satellites, which can detect such gamma rays from anywhere in the sky without any directed pointings.

The search for other electro-magnetic signals was aided by the localization of the source on the sky, made possible by triangulating the signals of the two LIGO detectors with constraints provided by a third detector called VIRGO in Italy. This led to a plethora of information about both the neutron stars and the associated material ejected from the merger, including spectroscopic signatures of very heavy elements such as silver, gold, and platinum. This supported earlier suggestions that such high-mass elements, colloquially characterized as "bling" because of their prominent use as

---

[3] See also the nice video at www.ligo.caltech.edu/video/ligo20160211v1.

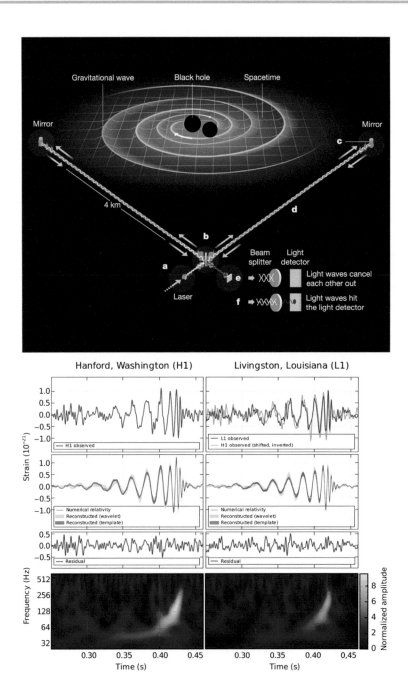

Figure 20.6 Top: Illustration for how the merger of orbiting black holes generates a spiral wave in spacetime, which upon propagation through the perpendicular arms of LIGO induces alternate stretches and compressions of the arm lengths that are detected by the interference pattern of two reflected laser beams. Bottom: Actual data traces from the first gravitational wave detection on September 17, 2015, as recorded from stations in Hanford, Washington, and Livingston, Louisiana. The lines compare the traces from the two stations with each other, and with numerical models; the bottom color plots show amplitude versus time and frequency, and the associated frequency increase that gives the characteristic "chirp" when translated into sound. Graphics courtesy Caltech/MIT/LIGO Laboratory.

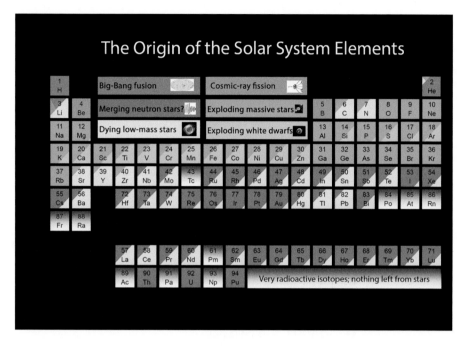

Figure 20.7 Periodic table showing the origin of each of the elements. The orange color shows that most of the heavy, neutron-rich elements, including precious "bling" like gold, platinum, and silver (Au, Pt, Ag), are now understood to have originated from the "kilonova" explosions from merging neutron stars. Graphic courtesy Jennifer Johnson, under Creative Commons Attribution-ShareAlike 4.0 International License. Astronomical image credits to ESA/NASA/AASNova.

precious metals and in jewelry, are mostly produced in the "kilonovae" associated with such neutron-star mergers, rather than, e.g., in core-collapse supernovae of massive stars, as had been previously widely believed. Figure 20.7 shows a periodic table with our current best estimates for the origin of each element.

Finally, while the energy flux of waves, either in light or gravity, declines with the inverse-square of distance of the source, the actual wave amplitude only drops *linearly* with inverse distance. Because LIGO directly detects this wave amplitude, each factor increase in the precision of the detection (from ongoing efforts to reduce the many sources of noise) leads to an equivalent factor increase in the distance that a source of given strength can be detected. But because the volume of space increases with the *cube* of this distance, the number of detectable systems increases with the cube of this improved precision. For example, a factor 2 increase in precision leads to a factor 2 in detectable distance, and so a factor 8 in the number of detectable sources. The expectation thus is that, with ongoing and planned improvements to sensitivity and precision, detection rates could approach one new source per day!

20.7   Questions and Exercises

Quick Questions

1. Derive a general formula for the escape speed v_{esc} for a neutron star of mass M. What is the value (in km/s) for $M = 2M_\odot$ and how does this compare to the Sun's escape speed, and to the speed of light?
2. What is the maximum escape speed (in km/s) of a neutron star?
3. Derive a general formula for log g of a neutron star of mass M, and compare this to log $g_\odot$, and also to log g_{wd} derived in Question 4 of Chapter 19?

Exercises

1. *Black-hole accretion disk*
   Consider accretion of mass onto a disk around a black hole of mass $3M_\odot$.
   a. What is the black hole's Schwarzschild radius R_{bh}, in km?
   b. What is the luminosity efficiency ϵ for accreting to a radius $3R_{bh}$ (correspond-ing to the lowest stable orbital radius in general relativity)?
   c. What is the accretion luminosity, in $L_\odot$, for a mass accretion rate $\dot{M}_a = 10^{-6} M_\odot/\text{yr}$ to this radius?
   d. Suppose this accretion luminosity is radiated as a blackbody from a disk ring with radii extending from 3 to 4 R_{bh}. Compute the ring's temperature T (in K). (Remember that a disk has *two* surfaces.)
   e. What is the wavelength λ_{max} (in nm) of peak flux for this radiation?
   f. What is the associated photon energy E (in eV)?
   g. What part of the electromagnetic spectrum does this correspond to?

2. *Disk emission*
   a. Use Wien's law to compute the peak wavelength (in nm) of thermal emission from the inner region of an accretion disk with temperature $T = 10^7$ K.
   b. What is the energy (in eV) of a photon with this wavelength?
   c. Now also answer both questions for $T = 10^{10}$ K.
   d. What parts of the electromagnetic spectrum do these photon wavelengths/energies correspond to?

3. *Pulsar spindown*
   Suppose that the period P of a pulsar increases such that $dP/dt = P/t_s$, where t_s is a constant, representing the characteristic "spindown" time for this pulsar.
   a. If the initial period at time $t = 0$ is P_o, derive an expression for $P(t)$.
   b. In terms of the pulsar's momentum of inertia $I \approx (2/5)MR^2$ and rotational frequency $\omega \equiv 2\pi/P$, next obtain the time evolution of rotational energy $E_{rot} = I\omega^2/2$.
   c. If the loss of rotational energy is emitted as radiation, derive then an expression for the associated luminosity $L_{rot}(t) = -dE_{rot}/dt$.

    d. Finally, evaluate L_{rot} (in $L_{\odot}$) for the present-day Crab pulsar, with $M = M_{\odot}$, $R = 10$ km, $P_o = 0.033s$, $t_s = 2500$ yr.

4. *Supernova ejecta shell into the ISM*

    Suppose a supernova explosion ejects a spherical shell of mass $M_e = 5M_{\odot}$ at an initial speed, $V_i = 10\,000$ km/s.

    a. Calculate the initial kinetic energy (in erg and J) of the shell, and also the total radial component of its momentum (in CGS and MKS units).

    b. Suppose the shell slows by sweeping up inter-stellar material, conserving this radial momentum. How much mass (in $M_{\odot}$) will be swept up when it has slowed to 10 km/s.

    c. If the inter-stellar medium is pure hydrogen with a density of 1 H-atom/cm^3, what is the radius of the shell when it reaches this speed of 10 km/s?

    d. Compare the kinetic energy of this 10 km/s shell with the initial energy computed in part a. Discuss what happened to any difference.

5. *Black-hole merger*

    Consider the merger of two black holes of equal mass $M_1 = M_2 = 30M_{\odot}$, leading to a final merged black hole of $M_f = 57M_{\odot}$.

    a. Compute the energy $E_{\Delta M}$ (in erg and J) associated with the mass deficit, $\Delta M = M_1 + M_2 - M_f$.

    b. What is each initial star's Scharzschild radius R_{bh}, in km?

    c. Using Kepler's laws, compute the orbital period P (in s and ms) when they are separated by a distance $3R_{bh}$ associated with the last stable orbit.

    d. Assuming the merger occurs over one such orbital period, compute the associated merger luminosity L_m, both in erg/s and $L_{\odot}$.

    e. Compare this with the total luminosity of all the stars in the universe, e.g., assuming $\sim 10^{11}$ galaxies with an average of $\sim 10^{11}$ stars, each with an average luminosity about that of the Sun.

Part III

Interstellar Medium and Formation of Stars and Planets

21 The Interstellar Medium

Compared with stars, the region between them, called the *interstellar medium* or "ISM," is very low density; but it is *not* a completely empty vacuum. For one thing, we have already seen that the final remnants of stars, whether white dwarfs, neutron stars, or black holes, generally have much less mass than the initial stellar mass; this implies that a substantial fraction (30 percent to 90 percent) of this initial mass is recycled back into the surrounding ISM through planetary nebulae, stellar winds, or supernova explosions. Indeed, a key theme in this Part III is that stars (and their planetary systems) are themselves *formed* out of this ISM material through gravitational contraction, making for a kind of star–gas–star cycle, as illustrated in Figure 21.1.

21.1 Star–Gas Cycle

If one assumes that, on average, a typical atom spends roughly equal fractions of time in the star versus the ISM phase of this cycle, then the average density of gas in the ISM should be roughly equal to the mass of the stars spread out over the volume between them. For example, in the region of the Galaxy near the Sun, the so-called solar neighborhood, the mean number density of stars is $n_* \approx 0.1 \, \mathrm{pc}^{-3}$, reflecting a typical interstellar separation distance $d \approx n_*^{-1/3} \approx 2$ pc. (Recall that the nearest star, Alpha, Centauri, is about 1.3 pc from the Sun.) If we take the average mass of each star to be roughly that of the Sun, we obtain a mean mass density $\rho \approx M_\odot n_* \approx 7 \times 10^{-24} \mathrm{g/cm}^3$.

This is much, much lower than the typical average density within stars, which as noted earlier for the Sun is $\rho_\odot \approx 1.4 \, \mathrm{g/cm}^3$; this mostly just reflects the huge distance/size ratio, giving roughly a factor $(\mathrm{pc}/R_\odot)^3 \sim 10^{24}$, for the volume between stars versus that within them. Thus while most stars typically have mean densities comparable to matter (such as water) here on Earth (i.e., $\rho \approx 1 \, \mathrm{g/cm}^3$), this ISM density is well below (by a factor $\sim 10^4$) even the most perfect vacuum ever created in terrestrial laboratories ($\rho \sim 10^{-19} \mathrm{g/cm}^3$).

Indeed, for ISM densities, it is more intuitive and common to quote values in terms of the *atom number density*. For example, with a composition dominated by hydrogen, the associated ISM hydrogen-atom number density is $n \approx \rho/m_\mathrm{p} \approx 4 \, \mathrm{cm}^{-3}$.

The assumptions and approximations behind this estimate – namely, the equal time between ISM and stars, the Sun representing a typical stellar mass, the assumed star

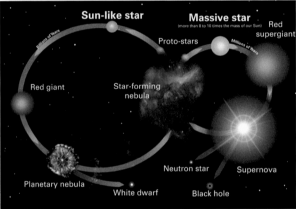

Figure 21.1 Top: "Pillars of creation" in the Eagle Nebula, a cold molecular cloud undergoing active star formation and illuminated by recently formed massive stars. Credit: NASA, ESA, and the Hubble Heritage Team (STScI/AURA). Bottom: Illustration of formation of solar-type and massive stars from interstellar cloud, showing the evolution toward end-stage remnants. Credit: NASA and the Night Sky Network.

density – are all somewhat rough, and can even vary significantly through the Galaxy. Nonetheless, when averaged over the entire Galaxy, the characteristic ISM number density $n \sim 1\,\mathrm{cm}^{-3}$ is indeed comparable to this local estimate. However, within this broad average, there are wide variations, reflecting a highly complex, heterogeneous, and dynamic ISM, as discussed next.

21.2   Cold–Warm–Hot Phases of Nearly Isobaric ISM

A key factor in the wide variation in density of the ISM is the wide variation in its temperature. Roughly speaking, gas in the ISM can be characterized in three

Figure 21.2 Maps along the plane of our Milky Way Galaxy, taken in multiple wavebands with energy increasing downward, ranging from radio continuum at the top, to gamma rays at the bottom. Each waveband is sensitive to distinct components of the multiphase temperatures of the ISM, along with the disk population of stars in the Galaxy. The map is oriented such that the center is toward the galactic center, in the direction of the constellation Sagittarius. Credit: NASA.

distinct temperature phases, ranging from cold ($T \sim 10-100$ K), to warm ($T \sim 5000-10,000$ K), to hot ($T \sim 10^5-10^7$ K). Figure 21.2 vividly illustrates the distinct signatures of these different components in various spectral wavebands ranging from the radio to gamma rays, as mapped along the disk plane of our Milky Way Galaxy.

In contrast to these wide variations in temperature, the gas *pressure*, which is proportional to the product of density and temperature, tends to be relatively constant over the broad ISM, with a typical value

$$P/k = nT \approx 10^3 \text{ K/cm}^3.$$ (21.1)

This near constancy of ISM pressure stems from the fact that gravity, which declines in strength with the inverse-square of the distance, is generally too weak to confine gas over the many parsec scales between stars.[1] In the absence of any restraining force, and ignoring any disturbances from stellar-mass ejection (e.g., from stellar winds or supernovae), the ISM gas should, over time, settle into a dynamical equilibrium that is roughly *isobaric*, meaning with a spatially constant gas pressure.

---

[1] An exception to this is in the densest cloud "cores" in star-forming regions, wherein gravity is compressing cold but very dense and thus high-pressure gas in the final contraction toward forming stars. See Chapter 22.

Within this roughly isobaric ISM, the densities of the three phases thus tend to scale inversely with temperature, ranging from $n \approx 10 - 100\,\mathrm{cm}^{-3}$ for cold clouds, to $n \approx 0.1 - 0.2\,\mathrm{cm}^{-3}$ for the warm gas, to $n = 10^{-2} - 10^{-3}\,\mathrm{cm}^{-3}$ for the hot component.

Because the flux of radiation also falls with the inverse-square of distance, we might expect the temperature of gas far from stars to be always very cold, for example as would be the case for the equilibrium temperature of a blackbody (see Exercise 1). But the low density of interstellar gas makes it very different from a blackbody, since emitting radiation requires collisions to excite atoms or interact with free electrons, the rates of which decrease with density.

Indeed, for the hot ISM the temperature is so high that hydrogen and helium are completely ionized, with only the heaviest and most complex atoms, such as iron, having a few remaining bound electrons. This means that radiative emission mostly occurs only through rare collisional excitation of these few, partially ionized heavy ions, making radiative cooling very inefficient, with a characteristic cooling time of several Myr. For the warm and still mostly ionized ISM, the higher density and greater number of bound electrons in heavy ions makes cooling somewhat more effective, but cooling times are still quite long, typically of order 10^4 yr.

As for the energy source that heats up the hot and warm ISM in the first place, this comes mainly from hot, massive stars. Near such a hot, luminous star, ultraviolet (UV) photoionization of the surrounding hydrogen gas can heat it to temperatures that are a significant fraction of the stellar effective temperature, of order 10^4 K. As detailed in Section 21.4, the resulting volume of ionized gas – dubbed HII regions from the standard notation for ionized hydrogen – can extend several parsecs from the star. Overall, such photoionization from hot stars is a significant heating source for the *warm* component of the ISM.

But an even more dramatic source of energy comes from the violent supernova (SN) explosions that end the relatively brief lifetimes of such hot, massive stars. In such SN explosions, several solar masses of stellar material are ejected at very high speeds, approaching 10 percent the speed of light, implying kinetic energies of order $E_{\mathrm{SN}} \sim M_\odot c^2 / 200 \sim 10^{52}$ erg. As this high-speed, expanding ejecta runs into the surrounding ISM, the resulting shock[2] wave heats the gas to very high temperatures, initially up to 10^8 K. But as the ISM gas piles up, the expansion slows and cools, ending up with a temperature $T \approx 10^5 - 10^7$ K, with the total pressure comparable to the surrounding ISM. Such SN explosions are thus the primary source of the *hot* component of the ISM.

Reflecting the large expansion of its source SN explosions, this hot ISM component can actually occupy more than half, even up to about 70 percent to 80 percent, of the *volume* of the Galaxy. Most of the remaining volume fraction, about 15 percent to 25 percent, makes up the warm component, with just a relatively small part, less than about 5 percent, being in relatively cool clouds.

---

[2] A shock wave arises whenever two gases collide with supersonic speed. It effectively converts the kinetic energy of the pre-shock gas into heat, yielding post-shock temperatures that scale with specific kinetic energy, or square of the speed, of the pre-shock gas.

And since thermal energy density is just $E_{th} = (3/2)nkT = (3/2)P$, the near constancy of ISM pressure means that the large volume of hot gas also contains most of the ISM thermal energy.

However, in terms of overall distribution of matter, most of the ISM *mass* is in relatively cool, dense clouds. As discussed in Chapter 22, it is these cool clouds that provide the source material for forming new stars, so let us next examine further their nature.

21.3   Molecules and Dust in Cold ISM: Giant Molecular Clouds

The low temperature and high density of the cold ISM makes it possible for the atoms to combine into molecules, and so much of the cold ISM takes the form of *giant molecular clouds* (GMCs). Reflecting the dominant abundance of hydrogen, the most common molecule is H_2. Among the heavy elements, carbon monoxide (CO) is usually the most abundant, reflecting its relative stability and the cosmic abundance of both its atomic constituents. Other common molecules include diatomic oxygen (O_2) and water (H_2O), and in some clouds up to 100 distinct molecular species (including, e.g., alcohol, CH_3CH_2OH) have been detected.

The survival, and thus abundance, of GMC molecules requires both a low local gas temperature and low UV flux, both of which become problematic in the vicinity of hot stars. But the high density and low temperature of such GMCs also means they tend to have quite high densities of ISM *dust*, and this can be very effective at shielding the regions from the heating and photodissociation by UV light from nearby stars. The dust itself is generally not formed locally, because in even the coldest clouds the density is not high enough for efficient nucleation of microscopic dust grains. Instead it is thought that most dust is formed in the outer layers of cool giant stars, and then blown away into the ISM by a strong stellar wind.

As discussed in Section 12.3, such dust can lead to very strong extinction and reddening of starlight. Figure 21.3 vividly illustrates the heavy extinction of the background starlight by the GMC Barnard 68. Note moreover how the partially extincted light from stars around the cloud edges is distinctly reddened.

To estimate the dust opacity, note that a spherical dust grain of radius a, mass m_d, and mass density $\rho_d = m_d/(4\pi/3)a^3$ has a physical cross section,

$$\sigma_d \equiv \pi a^2 = \frac{3m_d}{4a\rho_d}. \tag{21.2}$$

The overall opacity of a dust cloud is given by dividing this cross section for an individual dust particle by the mass m_c of cloud material *per* dust particle, i.e. $\kappa_d = \sigma_d/m_c$. For a mass fraction $X_d = m_d/m_c$ of a cloud that is converted into dust, we find, by using Eq. (21.2), that the implied opacity is

$$\kappa_d = \frac{3X_d}{4a\rho_d} \approx 150 \frac{cm^2}{g} \frac{X_d}{X_{d,\odot}} \frac{0.1\ \mu m}{a} \frac{1\ g/cm^3}{\rho_d}. \tag{21.3}$$

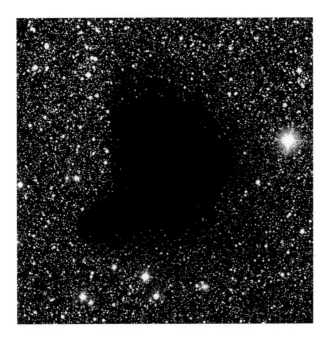

Figure 21.3 Color image of a star field surrounding the GMC Barnard 68, showing the heavy extinction of background starlight by its dust. Note in particular the reddening of partially extincted light around the cloud edges. Credit: ESO.

The latter equality provides a numerical evaluation scaled by: the dust mass fraction $X_{d,\odot} \approx 2 \times 10^{-3}$ assuming full conversion of dust-forming material at standard (solar) abundances; the dust grain internal density $\rho_d \sim 1\,\mathrm{g/cm^3}$ (most dust would almost float in water); and a typical dust grain size $a \approx 0.1\,\mu\mathrm{m}$. We thus see that the corresponding dust opacity, $\kappa_d \approx 150\,\mathrm{cm^2/g}$, is several hundred times greater than that for free electron scattering in the fully ionized gas inside a star, $\kappa_e \approx 0.34\,\mathrm{cm^2/g}$.

As already noted in Section 12.3, the opacity from the geometric cross section of dust grains applies only to wavelengths comparable to or smaller than the grain size, $\lambda \lesssim a$; for $\lambda > a$, the associated dust opacity decreases as $\kappa_d(\lambda) \sim (\lambda/a)^{-\beta}$, where the power index β is sometimes referred to as the "reddening exponent." For simple Rayleigh scattering from smooth spheres of fixed size a, $\beta = 4$. In practice, the complex mixtures in sizes and shapes of dust typically lead to a smaller effective power index, $\beta \approx 1{-}2$. Still, the overall inverse dependence on wavelength means that clouds that are optically thick to dust absorption and scattering will show a substantially *reddened* spectrum.

This reddening can be quantified in terms of a formal "color excess," which then can be used to estimate an associated visual extinction magnitude $A_V \equiv V_{obs} - V_{intrinsic}$. Recall from Section 12.3 that the extinction magnitude in any waveband is related to the associated optical depth, which in turns scales linearly with opacity in that waveband, $\kappa(\lambda)$. For wavelengths larger than the dust size $\lambda > a$, and assuming a linear reddening exponent $\beta \approx 1$, we thus see that the extinction magnitude declines with the inverse of the wavelength,

$$A(\lambda) \sim \tau(\lambda) \sim \kappa(\lambda) \sim 1/\lambda. \qquad (21.4)$$

For example, if a star has an extinction A_V in the visual waveband centered on $\lambda_V \approx 500$ nm, then in the mid-infrared (mid-IR) "M-waveband" at roughly a factor 10 higher wavelength $\lambda_M \approx 5000$ nm $= 5$ μm, the opacity, and thus the optical depth and extinction magnitude, are all *reduced* by this same factor 10, $A_M \approx A_V/10$. For a case with, say $A_V = 10.8$ magnitudes of visual extinction, the visual flux would be reduced by a factor $e^{-\tau_V} = e^{-A_V/1.08} = e^{-10} = 4.5 \times 10^{-5}$. By contrast, in this mid-IR M-band, the factor 10 lower extinction magnitude $A_M = 1.08$ implies a much weaker reduction, now just a factor $e^{-\tau_M} = e^{-A_M/1.08} = e^{-1} = 0.36$.

Stars are typically formed out of interstellar gas and dust in very dense molecular clouds, which often have 10 or 20 magnitudes of visual extinction ($A_V \approx 10 - 20$), essentially completely obscuring them at visual wavelengths. But such stars can nonetheless be readily observed with minimal extinction in mid-IR (a few microns) or far-IR (submillimeter) wavebands. As discussed in Chapter 13, this fact has spurred efforts to build large infrared (IR) telescopes, both on the ground and in space. The ground-based telescopes are placed at high altitudes of very dry deserts, to minimize the effect of IR absorption by water vapor in the Earth's atmosphere. Another issue is to keep the IR detectors very cold, to reduce their own thermal emission background.

Finally, the energy from dust-absorbed optical or UV light is generally reemitted in the mid-IR, at wavelengths set by the dust temperature through roughly the standard Wien's law for peak emission of a blackbody, $\lambda_{max} \approx 30$ μm$/(T/100$ K$)$ (cf. Eq. (4.6)). For GMC clouds with $T = 30-50$ K this gives thermal dust emission in the 60–100 μm range, as illustrated for galactic plane dust emission in Figure 21.2.

21.4 HII Regions

Let us next consider the warm ISM that is heated by UV photoionization from hot, luminous OB-type stars. Specifically, consider an ISM cloud with a uniform number density n of hydrogen atoms surrounding a hot star with luminosity L. Stellar UV photons with energy above the ionization energy of hydrogen, $h\nu > h\nu_0 \equiv 13.6$ eV, can efficiently ionize neutral hydrogen atoms; but these will then tend to quickly *recombine* with the free electrons.

The recombination rate scales with the electron density n_e times a temperature-dependent recombination coefficient,

$$\frac{1}{t_r} = n_e \langle \sigma_r v_e \rangle_T \approx 4 \times 10^{-13}(\text{cm}^3/\text{s}) n_e; \quad \text{for } T \approx 10^4 \text{ K}, \qquad (21.5)$$

where t_r is the recombination time for an ionized hydrogen atom, i.e., the time for a free proton to encounter and recombine with a free electrion. The recombination coefficient depends on the recombination cross section σ_r times the electron thermal speed v_e, with the angle brackets representing an average over the thermal velocity distribution of electrons at a temperature T. As mentioned above, UV photoionization tends to heat the gas to a temperature $T \approx 10^4$ K, and so the latter relation evaluates this recombination coefficient for that temperature.

The total emission rate of stellar UV ionizing photons can be estimated by integrating over the Planck blackbody function B_ν from the ionization threshold frequency ν_0, where $h\nu_0 \equiv 13.6\,\text{eV}$,

$$\dot{N}_{\text{UV}} \equiv L \int_{\nu_0}^{\infty} \frac{B_\nu\,d\nu}{B\,h\nu} = \frac{L}{Bh} \int_{\nu_0}^{\infty} \frac{B_\nu\,d\nu}{\nu}. \tag{21.6}$$

Here B is the spectrally integrated Planck function, and the division by the photon energy $h\nu$ converts the energy rate into a photon *number* rate.

In equilibrium, this number of ionizing photons will balance the total number of recombinations over a sphere (commonly dubbed a Strömgren sphere after the scientist who first described it) of radius R_S centered on the star. In terms of the proton (i.e., ionized hydrogen atom) number density n_p, each of the total number $n_p(4/3)\pi R_S^3$ of ionized hydrogen atoms in the sphere recombines with an electron of number density n_e over the recombination time t_r. The balance with stellar ionizing photons of emission rate $\dot{N}_{\text{UV}}$ thus requires

$$\dot{N}_{\text{UV}} = \frac{4\pi n_p R_S^3}{3t_r} = n_p n_e \, \langle \sigma_r v_e \rangle_T \, \frac{4\pi}{3} R_S^3. \tag{21.7}$$

For full ionization of a pure H cloud of number density n, we have $n_p = n_e = n$; thus the Strömgren radius R_S of such an "HII region" of ionized hydrogen (HII) is simply given by

$$R_S = \left[\frac{3\dot{N}_{\text{UV}}}{4\pi n^2 \, \langle \sigma_r v_e \rangle_T} \right]^{1/3} \approx 6.0\,\text{pc} \left[\frac{\dot{N}_{50}}{n_2^2} \right]^{1/3}, \tag{21.8}$$

where $\dot{N}_{50} \equiv \dot{N}_{\text{UV}}/(10^{50}\,\text{s}^{-1})$ and $n_2 \equiv n/(10^2\,\text{cm}^{-3})$ are convenient variables scaled by typical values for this photon rate and hydrogen number density.

The number of UV photons can be estimated from the spectral type of the exciting star, and the number density of hydrogen atoms can be inferred from the observed line emission from the HII region. By using these to compute the physical size R_S, we can then use the measured angular radius α to estimate the HII region's distance, $d = R_S/\alpha$.

In an HII region the ongoing recombination occurs through a cascade of electron transitions from higher to lower bound states of hydrogen, leading to extensive emission lines for all the hydrogen term series (Lyman, Balmer, Paschen, etc., for lower final state $n = 1, 2, 3$, etc.). But in optical images, the most prominent line emission stems from the $n = 3$ to $n = 2$ transition associated with the Balmer line Hα, which

Figure 21.4 Left: True-color optical image of the Rosetta Nebula and its associated HII region. The reddish glow is from Hα line emission from recombination of the ionized hydrogen. The central cavity has been evacuated by the strong, high-speed stellar winds from the central cluster of hot stars. Right: Composite false-color image showing the emission in Hα (red), and lines of OIII (green) and SII (blue). Credit: NASA/HST

is in the *red* part of the visible spectrum, with wavelength $\lambda = 656.28$ nm. Viewed in the visible, HII regions thus generally have a distinctly reddish glow, as illustrated in the left panel of Figure 21.4 for the HII region known as the Rosetta Nebula. But the false-color image of the same nebula in the right panel shows that, in addition to the Hα (now color-coded red), there is also line emission from doubly ionized oxygen (OIII, green) and singly ionized sulfur (SII, blue).

21.5   Galactic Organization and Star-Gas Interaction along Spiral Arms

In the dense regions of active star formation in the Milky Way and other galaxies, the ionization from numerous young, hot, massive stars can merge into extended *giant HII regions*. Viewed from Earth along the plane of the Milky Way, the projection of foreground and background stars and nebulae can make such regions appear complex and amorphous. A visually clearer view can be gleaned from external galaxies that are viewed *face on*, like the "Whirlpool galaxy" (M51) shown in Figure 21.5. The distinctly reddish splotches seen in the optical image in the left panel are all giant HII regions that formed in the dense clouds along this galaxy's spiral arms.

The right panel of Figure 21.5 shows a composite image in four distinct spectral bands, spanning the IR (red), optical (green), UV (blue), and X-rays (purple). The face-on view nicely complements the disk-embedded perspective images from multiple wavebands shown for own Milky Way galaxy in Figure 21.2. Note in particular how the close link between ISM and star formation is organized by the spiral arm structure. This is discussed further in the chapter on external galaxies (Chapter 27).

Figure 21.5 Left: Hubble Space Telescope optical image of M51, the "Whirlpool galaxy." The reddish blotches are from Balmer series Hα line emission (which at wavelength $\lambda = 656$ nm is in the red part of the visible spectrum) from giant HII regions. These represent the merger of many individual HII regions that arise when dense regions of interstellar hydrogen in otherwise cold giant molecular clouds (GMC) are photoionized by the UV radiation from the numerous, recently formed, hot massive stars. Note their proximity to dark bands formed from absorption of background stellar light by cold interstellar dust, which outline the galactic spiral arms. Right: Composite image of M51 from four NASA orbiting telescopes. X-rays (purple) detected by the Chandra X-ray Observatory reveal point-like sources from black holes and neutron stars in binary star systems, as well as a diffuse glow of hot ISM gas. Optical data from the Hubble Space Telescope (green) and infrared emission from the Spitzer Space Telescope (red) both highlight long lanes in the spiral arms that consist of stars and gas laced with dust. Finally, UV light (blue) from the GALEX telescope comes from hot, young stars, showing again how well these track the HII giants and star-forming GMCs along the spiral arms. Credit: NASA/HST/CXO/SST/GALEX.

21.6   Questions and Exercises

Quick Questions

1. What is the Strömgren radius R_S (in pc) of a medium with hydrogen number density $n_H = 280$ cm^{-3} around a hot star emitting $\dot{N}_{UV} = 8 \times 10^{50}$/s UV photons?
2. Suppose the above HII region is measured to have an angular diameter $\alpha = 1$ degree. What is its distance d (in pc)?
3. If dust in an ISM cloud has a reddening exponent $\beta = 2$, what is the ratio of its optical thickness between a wavelength λ and twice that wavelength?

Exercises

1. *Equilibrium temperature versus distance*
   a. Recalling that the equilibrium blackbody temperature of the Earth is $T_e \approx (T_\odot/2)\sqrt{2R_\odot/\text{au}} \approx 280$ K, show that the corresponding temperature at a distance d from the Sun is given by $T = T_e\sqrt{\text{au}/d}$.
   b. Compute the temperature for $d = 1$ pc.
   c. Compute the temperature at a distance $d = 1$ pc from a hot star with $T_* = 10\,T_\odot \approx 60\,000$ K.

    d. How do these temperatures compare to the hot, warm, and cold phases of the ISM?

2. *Hα luminosity of HII region*

   A star emits UV photons at a rate $\dot{N}_{UV} = 8 \times 10^{50}$/s.

    a. What is the luminosity L (in erg/s and $L_\odot$) of the associated HII region in Hα if every recombining hydrogen atom goes through the $n = 3$ to $n = 2$ transition?

    b. How would this change if only fraction f of recombining atoms went through that transition?

    c. How does this depend on the local density of hydrogen, and on the associated Strömgren radius?

3. *HII region scalings*

   Consider a star with temperature $30\,000$ K (about $5 \times T_\odot$) and radius $R = 40\,R_\odot$.

    a. Compute the total luminosity L in $L_\odot$.

    b. For the hydrogen ionization energy $E = h\nu_0 = 13.6$ eV, compute the ratio $h\nu_0/kT$.

    c. Using the fact that $h\nu_0/kT \gg 1$, one can show that the dimensionless ratio, $b \equiv \dot{N}_{UV}h\nu_0/L \approx 0.17$, where $\dot{N}_{UV}$ is the number of hydrogen-ionization UV photons (defined in Eq. (21.6)). Use this to compute a numerical value for $\dot{N}_{UV}$ (in photons/s.).

    d. Assuming an interstellar hydrogen density $n = 8000$ cm^{-3}, compute the associated equilibrium (Strömgren) radius R_S of the resulting HII region (in pc).

    e. If such an HII region has an apparent angular diameter of 1 arcmin, use this and the answer to part d to determine the distance d (in kpc).

    f. Using the approximation $h\nu_0/kT \gg 1$ and the definitions of the Planck function and $\dot{N}_{UV}$, compute from first principals the value of the ratio b in part c, confirming the value given there.

4. *GMC dust extinction and reddening*

    a. For a GMC with molecular hydrogen density $n = 100$ cm^{-3}, compute the associated mass density ρ.

    b. For UV light with $\lambda = 100$ nm and dust with size $a = 0.1$ μm and solid density $\rho_d = 1$g/cm^3 and the solar abundance mass fraction $X_d = 2 \times 10^{-3}$, use the geometric dust opacity derived in the text to compute the mean free path ℓ (in pc) for this GMC.

    c. For a GMC of diameter $D = 30$ pc, compute the optical depth τ and reduction fraction F_{obs}/F for a star behind the cloud that emits such UV light.

    d. Use this to compute the associated extinction magnitude for this UV light, A_{UV}.

    e. Assuming a reddening exponent $\beta = 1$, now compute the extinction A_V for visible light with $\lambda = 500$ nm, and the extinction A_{NIR} for near IR light with $\lambda = 2$ μm.

22   Star Formation

22.1   Jeans Criterion for Gravitational Contraction

Stars generally form in clusters from the gravitational contraction of a dense, cold GMC. The requirements for such gravitational contraction depend on the relative magnitudes of the total internal thermal (kinetic) energy K versus the gravitational binding energy U. For a cloud of mass M, uniform temperature T, and mean mass per particle μ, the total number of particles $N = M/\mu$ have an associated total thermal energy,

$$K = \frac{3}{2} N k T = \frac{3}{2} \frac{M k T}{\mu}. \qquad (22.1)$$

If the cloud is spherical with radius R and uniform density $\rho = \mu n = M/(4\pi R^3/3)$, the associated gravitational binding energy (cf. Eq. (8.3)) is

$$U = -\frac{3}{5} \frac{G M^2}{R}. \qquad (22.2)$$

Recalling the condition $K = -U/2$ for stably bound systems in *virial* equilibrium, we can expect that for a cloud with $K > -U/2$, the excess internal pressure would do work to expand the cloud against gravity, leading it to to be less tightly bound (or even unbound, if $K > -U$).

Conversely, for $K < -U/2$, the too-low pressure would allow the cloud to gravitationally contract, leading to a more strongly bound cloud. The critical requirement, known as the *Jeans criterion*, for such gravitational contraction can thus be written

$$\boxed{\frac{M}{R} > \frac{5kT}{G\mu}}. \qquad (22.3)$$

In terms of the cloud's atomic number density $n = \rho/\mu = N/(4\pi R^3/3)$, we can define a minimal *Jeans radius* for cloud contraction,

$$\boxed{R_{\mathrm{J}}} \approx \left(\frac{15kT}{4\pi n G \mu^2} \right)^{1/2} \boxed{\approx 9.6\,\mathrm{pc} \left(\frac{T}{n} \right)^{1/2} \frac{m_{\mathrm{p}}}{\mu}}, \qquad (22.4)$$

where the second equality assumes CGS units, with number density n in cm^{-3} and temperature T in kelvin.

Alternatively, one can define a minimum *Jeans mass* (the total mass within a Jeans radius) for a cloud to contract,

$$\boxed{M_J} \equiv \frac{4\pi R_J^3}{3}\,\mu n \approx \frac{5}{\mu^2}\left(\frac{kT}{G}\right)^{3/2}\left(\frac{15}{4\pi n}\right)^{1/2} \boxed{\approx 92\,M_\odot\frac{T^{3/2}}{n^{1/2}}\left(\frac{m_p}{\mu}\right)^2}. \quad (22.5)$$

For typical ISM conditions, both the Jeans radius and mass are quite large, implying it can be actually quite difficult to initiate gravitational contraction. For example, for a cold cloud with $\mu = m_p$, $T = 100\,\text{K}$ and $n = 10\,\text{cm}^{-3}$, we find $R_J \approx 30\,\text{pc}$ and $M_J \approx 30\,000\,M_\odot$, requiring a cloud that is initially extremely large and massive. The requirements are somewhat less severe once the hydrogen atoms form into H_2 molecules, thus increasing the molecular weight to $\mu \approx 2m_p$, and so reducing R_J by a factor 2, and M_J by a factor 4.

But a general upshot of such a large Jeans mass is that stars tend typically to be formed in large clusters, resulting from an initial contraction of a GMC, with mass of order $10^4\,M_\odot$ or more.

22.2   Cooling by Molecular Emission

In contrast to the poor radiative efficiency of the ionized gas in the warm and hot phases of the ISM, in the cool ISM the formation of molecules makes such clouds much more efficient for radiative cooling. The thermal, collisional excitation of the molecules and dust leads to emission of radiation at IR wavelengths comparable to those associated with blackbody emission for the given temperature. For example, for a cloud with temperature $T = 100\,\text{K}$, radiation is at IR wavelengths $\lambda \approx \lambda_{\text{max},\odot}T_\odot/T \approx 30\,\mu\text{m}$ (see Eq. (4.6)).

At low temperatures $T < 100\,\text{K}$, cooling by molecular radiation is dominated by carbon monoxide (CO). Both carbon and oxygen are relatively abundant elements, and the molecular structure of CO provides a variety of excitation modes (rotational, vibrational, or electronic) from inelastic collision with molecular hydrogen. This converts kinetic energy of the gas to potential energy in the molecules, which deexcite radiatively to emit an IR photon that escapes the cloud, causing it to cool.

Such CO molecular cooling is a key factor in initiating and maintaining cloud contraction, by allowing the cloud to shed the increased internal energy gained from the tighter gravitational binding. In virial equilibrium only half this energy is lost, and so the interior would still heat up in proportion to the stronger gravitational binding. But in practice, CO emission is often so efficient that the cloud interior can stay cool, or even become cooler, as it contracts. The resulting dramatic reduction in interior pressure support then leads to a full gravitational collapse.

22.3   Free-Fall Timescale and the Galactic Star-Formation Rate

To estimate the timescale for gravitational collapse, recall first from Kepler's third law (see Eq. (10.5)) that the period P for orbit at radius R around an object of mass M is

$$P = \sqrt{\frac{4\pi^2 R^3}{GM}} = \sqrt{\frac{3\pi}{G\rho}}, \tag{22.6}$$

where the second equality casts this in terms of the mean density within a sphere of this radius, $\rho = M/(4\pi R^3/3)$. Since the period from Kepler's law does not depend on the orbital eccentricity, Eq. (22.6) also applies to a purely radial orbit (with eccentricity $\epsilon = 1$) through a central point mass from this radius. But the self-gravitational collapse of a cloud would occur at just this same rate, implying then that the free-fall time to contract to zero radius from the initial radius R should be just a quarter of this orbital period,

$$\boxed{t_{\mathrm{ff}} = \frac{P}{4} = \sqrt{\frac{3\pi}{16G\rho}} = \frac{0.82\,\mathrm{hr}}{\sqrt{\rho}} = \frac{51\,\mathrm{Myr}}{\sqrt{n}}\sqrt{\frac{2m_{\mathrm{p}}}{\mu}},} \tag{22.7}$$

where the density evaluations assume CGS units (i.e., g/cm^3 for ρ and cm^{-3} for n). For a star such as the Sun, for which the mass density is CGS order unity, free-fall would be less than an hour. But for a cold molecular cloud, with say a number density $n \approx 100\,\mathrm{cm}^{-3}$, such free-fall would take several Myr.

In our Galaxy, the total mass in giant molecular clouds with density $n \gtrsim 100\,\mathrm{cm}^{-3}$ is estimated to be about $M_{\mathrm{GMC}} \approx 10^9 M_{\odot}$. Since this mass should collapse to stars over a free-fall time, it suggests an overall galactic star-formation rate should be given by

$$\dot{M}_{\mathrm{sfr}} = \frac{M_{\mathrm{GMC}}}{t_{\mathrm{ff}}} \approx 200 \frac{M_{\odot}}{\mathrm{yr}}. \tag{22.8}$$

But the observationally inferred star-formation rate is actually much smaller, only about 1 $M_{\odot}$/yr, implying an effective efficiency of only $\epsilon_{\mathrm{ff}} \lesssim 0.01$. The reasons for this are not entirely clear, but may stem in part from inhibition of gravitational collapse by interstellar magnetic fields, and/or by interstellar turbulence. Another likely factor is the feedback from hot, massive stars, which heat up and ionize the cloud out of which they form, thus inhibiting the further gravitational contraction of the cloud into more stars.

22.4   Fragmentation into Cold Cores and the Initial Mass Function

In those portions of a GMC that do undergo gravitational collapse, the contraction soon leads to higher densities, and thus to smaller Jeans mass and Jeans radius, along with a shorter free-fall time. This tends to cause the overall cloud, with total mass

$10^4 - 10^6 M_\odot$, to fragment into much smaller, stellar-mass cloud "cores" that will form into individual stars.

A key, still-unsolved issue in star formation regards the physical processes and conditions that determine the mass distribution of these proto-stellar cores, leading then to what is known as the stellar *initial mass function* (IMF).

This IMF can be written as dN/dm, wherein $m = M/M_\odot$ is the stellar mass in solar units, and $dN(m) = (dN/dm)dm$ represents the fractional number of stars within a mass range m to $m + dm$. Studies of the evolution of stellar clusters suggest that this can be roughly characterized by a power-law form,

$$\boxed{\frac{dN}{dm} = Km^{-\alpha},}$$
(22.9)

where K is a normalization factor that depends on the total number of stars, and the power index α has distinct values for high-mass versus low-mass stars. For $m > 1$, the most commonly inferred value is $\alpha \approx 2.35$, known as the "Salpeter" IMF (solid red line in Figure 22.1), after the scientist who first quantified the concept of an IMF. The large power-index reflects the fact that higher-mass stars are much rarer than lower-mass stars.

More modern models generally flatten the distribution at lower mass, as illustrated in Figure 22.1. This allows them to be normalized to a finite number (set to unity in Figure 22.1) when integrated over all masses.

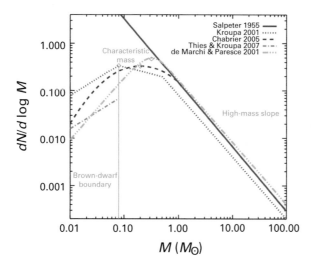

Figure 22.1 Comparison of IMFs from various authors, plotted on a log–log scale in terms of the mass fraction per logarithm interval of mass, which is proportional to $m\,dN/dm$. Except for the Salpeter pure-power-law form, the curves are normalized such that the integral over mass is unity. From review by Offner et al., in *Protostars and Planets VI*, University of Arizona Press (2014), eds. H. Beuther, R. S. Klessen, C. P. Dullemond, Th. Henning.

With a given form of the IMF for a collapsing GMC, one can model the evolution of the resulting stellar cluster, based on how each star with a given mass evolves through its various evolutionary phases, e.g., main sequence, red giant, etc.

22.5   Angular Momentum Conservation and Disk Formation

In general, the fragmentation of a GMC into stellar-mass cores will endow those cores with a nonzero rotation, and this can be a key factor in their final collapse toward stellar size. While material near and along the core rotation axis can still collapse to form the central star, the conservation of angular momentum for material near the rotational equator can halt the contraction and lead to formation of a *proto-stellar disk*.

For material with angular momentum per unit mass $j \equiv vr$ in circular orbit with speed v at a radius r about a central mass M, the orbital condition for balance between centrifugal and gravitational acceleration can be cast in the form

$$\frac{GM}{r^2} = \frac{v^2}{r} = \frac{j^2}{r^3}. \tag{22.10}$$

For an initially spherical core with starting radius R and angular rotation frequency Ω, a mass parcel at the rotational equator has an angular momentum per unit mass $j_{eq} = \Omega R^2$. As the cloud collapses under the gravitational attraction of its own mass M, conservation of angular momentum causes this parcel to rotate faster until it reaches the condition (22.10) for orbit, with an associated "disk" radius,

$$r_d = \frac{j_{eq}^2}{GM} = \frac{\Omega^2 R^4}{GM} \equiv 2\beta_{eq} R. \tag{22.11}$$

The last equality here introduces the initial equatorial ratio of rotational to gravitational energy,

$$\boxed{\beta_{eq} \equiv \frac{\Omega^2 R^2/2}{GM/R} = \frac{3\Omega^2}{8\pi G\rho} = \frac{\Omega^2 P_{orb}^2}{8\pi^2} = \frac{1}{2}\frac{\Omega^2}{\Omega_{orb}^2}.} \tag{22.12}$$

The second equality here shows that β_{eq} depends only on the core density ρ and its rotation frequency Ω, two quantities that can generally be readily inferred from observations, with observed cloud cores typically giving $\beta_{eq} \approx 0.02$. For a typical observed core size $R \approx 0.05\,\mathrm{pc}$, the expected disk radius r_d is a few hundred au, comparable to the inferred sizes of proto-stellar disks.

The last two equalities recast β_{eq} in terms of the orbital period P_{orb} or orbital frequency $\Omega_{orb} \equiv 2\pi/P_{orb}$. For various assumed values of β_{eq}, Figure 22.2 illustrates how material from various latitudes in the original clouds track toward the center star or disk under gravitational collapse. For a cloud with initial radius R and uniform density ρ, Figure 22.3 plots the radial variation of the resulting disk's surface density Σ (in units of ρR), for the labeled values of equatorial rotation parameter β_{eq}.

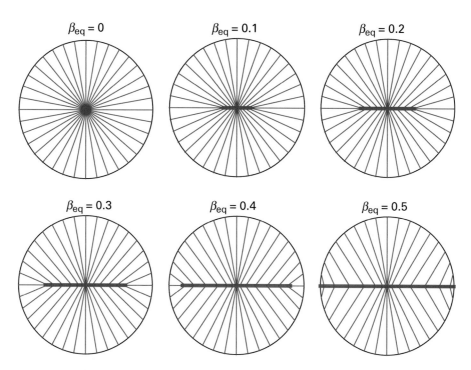

Figure 22.2 Traces (blue lines) illustrating how conservation of angular momentum causes various locations on the surface of a rigidly rotating spherical cloud (represented by black circle) to collapse onto an orbiting disk (marked in red). The various panels are for the labeled values β_{eq} of the equatorial rotational energy to gravitational energy. Note how material near the rotational poles contracts to the concentrated central region, while material at lower latitudes near the equator collapses onto the orbiting disk with outer radius $r_d = 2\beta_{eq}R$, as given by Eq. (22.11).

Initially such disks can have a mass that is a substantial fraction (about 35 percent) of that for the central star. But in disks with Keplerian orbits, the orbital frequency increases inward with radius as $\Omega_{orb} \sim r^{-3/2}$, meaning that between two neighboring rings there is an overall *shear* in orbital speed. Any frictional interaction – e.g., due to viscosity – between such neighboring rings will thus tend to transport angular momentum from the faster inner ring to the slower outer ring, allowing the inner mass to fall further inward, while the angular momentum receiving material moves further outward. Because the specific angular momentum increases outward as $j = vr = \Omega r^2 \sim \sqrt{r}$, this outward viscous diffusion of angular momentum allows, over time, for most of the mass to accrete onto the star, with just a small mass fraction retaining the original angular momentum.

Eventually this remnant disk-mass can fragment into its own gravitationally collapsing cores to form planets. In our own solar system the most massive planet Jupiter has only 0.1 percent the mass of the Sun, but 99 percent of the solar system's angular momentum. Of course, Earth too originated from the evolving proto-solar disk. You and I and everyone on Earth are here today because our source material happened

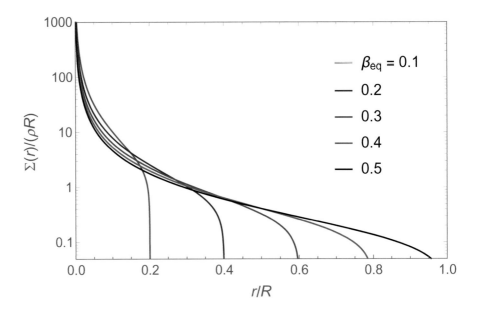

Figure 22.3 Disk *surface density* $\Sigma(r)$ (i.e., mass per unit *area* of the disk) vs. disk radius r for the simple model of the gravitational collapse of a rotating, initially spherical cloud. The curves show results for various initial equatorial ratios of the rotational to gravitational energy, $\beta_{eq} = 0.1$–0.5. The disk surface density here is in units of ρR, where ρ and R are the cloud's initial mass density and radius.

to stem from the equatorial regions of the proto-solar core, with too much angular momentum to fall into the Sun itself; it also then could have been the viscous recipient of the angular momentum from other proto-solar-disk material that did diffuse inward onto the Sun. The formation of such planetary systems around our Sun and other stars is discussed next, in Chapter 23.

22.6   Questions and Exercises

Quick Questions

1. A cold cloud of molecular hydrogen has a temperature $T = 40$ K. Assuming it has the standard ISM equilibrium pressure value for nT, what is the associated Jean's mass $M_{\rm J}$, in $M_{\odot}$?

2. a. What is the free-fall time for a star such as the Sun?

   b. What is the free-fall time for a GMC with number density $n = 100\,{\rm cm}^{-3}$ of molecular hydrogen?

3. *Free-fall time for simple constant-gravity model*

   Recall the elementary physics result that an object falling under gravitational acceleration g drops a distance $s = gt^2/2$ in time t.

a. Fixing the gravity at a constant value $g = GM/R^2$, use this simple relation to solve for the time $t_g(R)$ to fall through a stellar radius (i.e., by setting $s = R$).

b. Compare this with the free-fall time in Eq. (22.7) by evaluating the ratio $t_g(R)/t_{ff}$.

Exercises

1. *Disk collapse from various latitudes*
   a. For a spherical cloud of radius R and rotation frequency Ω, consider locations away from the equator, with colatitude θ measured from the polar axis. Derive an expression for the associated ratio $\beta(\theta)$ of the local rotational energy to gravitational energy, writing this in terms of the equatorial ratio β_{eq} derived in Eq. (22.12).
   b. Use this to derive an expression for the associated disk radius $r(\theta)$ to which material contracts from various latitudes on the initial spherical surface of radius R. (You may assume that throughout the contraction, the gravitational attraction is that from a point source of mass M at the cloud center.) The blue lines in Figure 22.2 draw connections between this disk radius and its source location at various latitudes on the cloud surface, for various choices of the parameter β_{eq}.

2. *Isobaric ISM*
   a. Assuming an isobaric ISM with the canonical pressure $P/k = nT = 10^3 \, \text{K cm}^{-3}$, derive expressions for R_J (in pc) and M_J (in $M_\odot$) as a function of temperature T (in K).
   b. Now derive analogous expressions for R_J and M_J as a functions of number density n (in cm^{-3}).
   c. Assuming the molecular weight $\mu = 2m_p$ for pure molecular hydrogen, what would be the density n (in cm^{-3}) of a Jean's cloud with this pressure and a mass $M = 1000M_\odot$?
   d. What would be the associated Jean's radius R_J (in pc)?
   e. What angle (in arcsec) would the cloud *diameter* subtend at a distance of 1 kpc from the Earth?
   f. Finally, what would be the free-fall time t_{ff} (in years)?

3. *Flattened Salpeter IMF*
   a. For the simple flattened Salpeter IMF, with $\alpha = 2.35$ for $m > 1$ and $\alpha = 0$ for $m < 1$, integrate Eq. (22.9) over all masses to obtain an expression for the normalization K in terms of the total number of stars N_{tot}.
   b. Now use this to obtain an expression for the *fraction* of stars, $N(m > m_o)/N_{tot}$, with mass greater than some mass lower limit m_o (assuming $m_o \geq 1$). In particular, what fraction of stars have $m > 1$?
   c. For $m_o = 100$, how many total stars must a cluster have for there to be at least one star with $m \geq m_o$? How about for $m_o = 300$? What does this imply for observational efforts to determine whether there is an upper mass cutoff to the IMF?

Challenge Exercise

4. *Disk surface density*

   Consider a *hollow* thin spherical shell of radius R and rotation frequency Ω that collapses under the gravitational attraction of a star of mass M_* at the shell center.

   a. Assuming the shell has a mass M_s that is initially spread uniformly over its spherical surface, use the results of the previous exercise to derive an expression for the disk surface density $\Sigma(r)$ as a function of disk radius r. Express this in terms of the shell mass M_s, the outer disk radius r_d in Eq. (22.11), and the ratio r/r_d.

   b. Now use this result to derive an integral expression for the total disk surface density $\Sigma(r)$ from collapse of a *filled*, constant-density, spherical cloud of radius R, mass M, and (rigid-body) rotation frequency Ω. (You may assume that the mass $M(r)$ inside any material initially at radius $r \leq R$ remains constant throughout the contraction.) Figure 22.3 plots results for such a disk model.

23 Origin of Planetary Systems

23.1 The Nebular Model

The disk-formation process of the previous chapter forms the basis for the "nebular model" for the formation of planetary systems, including our own solar system. As illustrated in Figure 23.1, as a proto-stellar cloud collapses under the pull of its own gravity, conservation of its initial angular momentum leads naturally to formation of an orbiting disk, which surrounds the central core mass that forms the developing star. With the usual interstellar composition of mostly hydrogen and helium, and only about 1 percent to 2 percent of heavier elements, this disk is initially gaseous, held in a vertical hydrostatic equilibrium about the disk mid-plane, with the radial support against stellar gravity provided by the centrifugal force of its orbital motion.

This stops the rapid, dynamical infall, but as the viscous coupling between differentially rotating rings transports angular momentum outward, there remains a relatively slow inward diffusion of material that causes much of the initial disk mass to gradually accrete onto the young star. This, along with other effects – such as the entrainment of disk material by an outflowing stellar wind – gradually depletes the hydrogen and helium gas in the disk. But during this slow dissipation of the disk mass, which likely occurs over a few million years, the heavier trace elements can, in the relatively dense conditions of this disk, gradually bond together to make molecules. These in turn nucleate into ever-growing grains of dust, and eventually into rocks and even boulders.

Collisions among these boulders leads to a combination of fragmentation and accumulation, with the latter eventually forming asteroid-size (meters to kilometers) bodies for which self-gravity becomes significant. This first forms loosely held "rock piles," then planetoids, and eventually planets. The detailed process are chaotic, with frequent collisions, but eventually, through accretion and assimilation, the largest bodies clear out most of the debris that shared their orbital distance from the central star.

23.2 Observations of Proto-Planetary Disks

While the basic ideas behind the nebular model date back to Kant and Laplace in the eighteenth century, modern observations now provide direct support for the overall model, and increasingly strong constraints on its specifics. Young stellar objects (YSOs), identified spectroscopically to still be contracting to the main sequence along

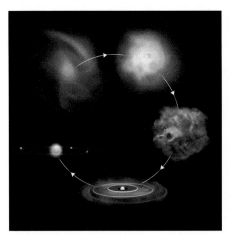 

Figure 23.1 Left: Illustration of the nebular model for formation of a planetary system. Credit: Bill Saxton/NRAO/AUI/NSF. Right: Direct image of proto-planetary disk in the T Tauri star HL Tauri, made in millimeter wavelengths with the Atacama Large Millimeter Array (ALMA). The entire disk spans about 200 au. The disk gaps likely represent regions of planet formation. Credit: ALMA (ESO/NAOJ/NRAO), NSF.

the Hayashi or Henyey tracks (Section 17.4), often show clear evidence of proto-planetary disks. Herbig Ae/Be stars are relatively hot, massive YSOs that show strong hydrogen emission from a gaseous disk. The cooler, lower-mass T Tauri stars often show an infrared excess thought to arise from dust thermal emission in a warm proto-planetary disk.

With advent of telescope arrays (e.g., ALMA, see Section 13.2) observing in the far infrared (IR) and submillimeter spectral regions, it is now becoming possible to image such disks directly. The right panel of Figure 23.1 shows an ALMA image of a proto-planetary disk in the T Tauri star HL Tauri, made in millimeter wavelengths. The star's temperature and luminosity put it on the Hayashi track of the H–R diagram, in a pre-main-sequence phase with age less than 1 Myr. Interferometry from the array allows spatial resolution ranging down to 0.025 arcsec. At HL Tauri's distance of 140 pc, this corresponds to 3.5 au, with the visible disk extending over a diameter of about 200 au. The disk gaps likely represent regions where planet formation is clearing out disk debris, though there is so far no direct evidence of fully formed planets in this system.

Disks similar to that around HL Tauri have been inferred around other very young stars, but with densities that generally degrade with stellar age, over timescales of a few Myr. The Hubble Space Telescope has imaged several "debris disks" around stars with ages ~10 Myr. These are thought to be the later stages when the disk debris has been depleted by various processes, like accretion onto the star, dissociation by stellar UV radiation, and entrainment in a outflowing stellar wind. A key issue in planet formation is thus whether this can occur quickly enough to compete with such disk depletion.

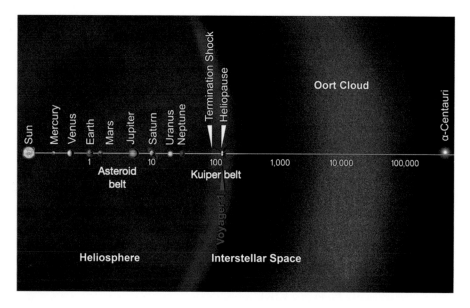

Figure 23.2 Illustration of key components of our solar system, extended on a log scale into the local ISM, through the Oort cloud, and to the nearest star. Credit: NASA.

23.3   Our Solar System

The above nebular model for formation of planetary systems implies that planets should be quite common around stars with less than a few solar masses. Indeed, the techniques that have led to positive detection of some 4000+ such extra-solar planets ("exoplanets") are discussed in Chapter 25. But our own solar system still provides a key, best-observed prototype, and so as background, let us first briefly review the key properties of the bodies and material that surround and orbit our Sun (see Figure 23.2).

Chief among these are the eight planets,[1] which can be quite conveniently divided into the four relatively small, rocky inner planets (Mercury, Venus, Earth, Mars), and the four outer giants made of mostly gas (Jupiter and Saturn) and ice (Uranus and Neptune). They all have roughly circular orbits that lie in nearly the same plane as the orbit of the Earth, known as the *ecliptic*. Most have rotations that, with various tilts, are in the same sense as their orbit, the exceptions being Venus, which rotates slowly backward, and Uranus, whose rotational axis nearly lies in the plane of its orbit.

Between the orbits of Mars and Jupiter, there is a belt of smaller *asteroids* that likewise mostly have nearly circular orbits in this ecliptic plane. Most are small ($\sim$1 m to 1 km) with irregular shapes, but the largest, Ceres, with radius $R \approx 1200$ km, is massive enough to be made spherical by its self-gravity, and so is classified a "dwarf planet." Asteroid orbits are strongly influenced by Jupiter's strong gravity, which is systematically clearing the region. Their combined mass is estimated to be only about 4 percent that of Earth's Moon.

---

[1] The erstwhile ninth planet, Pluto, has now be reclassified as a "dwarf planet," and part of the Kuiper Belt discussed below.

The gas giants all have multiple satellites, formed by a smaller-scale version of angular momentum conservation as their proto-planetary clouds were contracted by gravity. This left Jupiter and Saturn with several moons that, while much smaller than their host planet, have comparable size to Earth's Moon. These include Jupiter's four "Galilean" moons[2] (Io, Europa, Ganymede, and Callisto), and Saturn's Titan, the only moon with a dense atmosphere, composed largely of nitrogen and methane. Saturn also has its prominent ring systems, which are composed of a large number of icy bodies ranging from centimeters to several meters across. The other gas giants have thinner, weaker rings. Jupiter's moon Europa also shows evidence for extensive surface ice, as well as a subsurface ocean.

Somewhat beyond the orbit of Neptune are icy "Kuiper Belt Objects" (KBOs), which have somewhat eccentric and inclined orbits in a belt at distance 30–50 au. Originally predicted based on theoretical arguments, there are now more than 2000 directly detected, with more than 100 000 thought to exist with diameter >100 km. A few, including Pluto and its companion Charon, are large enough to also be considered dwarf planets. Indeed, their discovery was a major factor in the decision to reclassify Pluto as a KBO.

At much greater distances (>1000 au) lies the "Oort cloud" of icy planetesimals, with eccentric orbits that flare from near the ecliptic in the inner regions, to a nearly spherical distribution in the outer cloud. When deflected into the inner solar system, heating of the ice by the Sun causes outgassing, which with the outward push from radiation and the solar wind form the characteristic comet tails.

23.4   The Ice Line: Gas Giants versus Rocky Dwarfs

For the *gas giants* (Jupiter and Saturn) and *ice giants* (Uranus and Neptune) in the outer solar system, we need also to consider the role the much cooler conditions in allowing the formation of water ice.

As shown in Figure 6.3, hydrogen and oxygen are the first and third most abundant elements in the Sun. In the solar nebula, their ready combination into water molecules (H_2O) thus made that relatively abundant. In the colder outer regions these condensed to form ice, which gradually collected into ever larger solid cores, eventually growing massive enough to gravitationally attract and retain the even more abundant but lighter gases of hydrogen and helium. This was the basis for formation of the outer gas and ice giant planets, with an overall composition similar to the solar nebula, and the present-day Sun. In contrast, in the inner nebula, where it was too warm to form ice, such light atoms of hydrogen and helium escaped from the weaker gravity of the smaller, rocky planets, effectively stunting their growth and so keeping them relatively small.

To quantify this "ice line" between inner rocky dwarfs and outer gas giants, let us next derive an estimate for the decline of equilibrium temperature with distance from the Sun.

---

[2] So named because they were first discovered by Galileo.

23.5   Equilibrium Temperature

For an absorbing sphere with radius r at a distance d from the Sun, the intercepted flux of the Sun's luminosity $L_\odot$ is $\pi r^2 L_\odot/4\pi d^2 = \pi r^2 \sigma_{sb} T_\odot^4 (R_\odot/d)^2$, with $R_\odot$ and $T_\odot$ (≈ 5800 K) the Sun's radius and effective temperature, and σ_{sb} the Stefan–Boltzmann constant (see Section 5.2). If we assume this sphere then radiates this energy as a blackbody over its surface area, $4\pi r^2 \sigma_{sb} T^4$, then solving for its equilibrium temperature gives

$$T(d) = T_\odot \sqrt{\frac{R_\odot}{2d}} \approx 290\,\text{K}\,\sqrt{\frac{\text{au}}{d}}. \tag{23.1}$$

Note here that the sphere's radius r has cancelled, and so in principle this could be applied to bodies of any size, ranging from grains of dust to whole planets.

Indeed, for an object orbiting at the Earth's distance of 1 au, the equilibrium temperature of about 290 K (i.e., just above water's freezing point of 273 K) is pretty close to the actual mean temperature of Earth itself. But that is the result of a somewhat fortuitous and delicate cancellation, between the cooling effect of reflection of sunlight by clouds, and the warming effect of greenhouse gases in the Earth's atmosphere.

By contrast, at a distance of about 0.7 au, the planet Venus would, by Eq. (23.1), be predicted to have a temperature just about 15 percent higher than Earth, ~ 350 K; but in fact, due to a *runaway* greenhouse effect, the surface temperature on Venus is more than twice this value, >700 K.

On the other side, for its distance at about 1.5 au, Mars has a predicted equilibrium temperature ~ 235 K. Owing to the lack of much greenhouse effect from its much thinner atmosphere, this is pretty close to Mars' actual average surface temperature. While there is much evidence that Mars once had a much thicker atmosphere, and a warm enough temperature to have had liquid water flowing across its surface, today all its water is locked up in ice, at its poles and below its surface.

23.6   Questions and Exercises

Quick Questions

1. For material in a proto-stellar Keplerian disk around a star with mass M, derive formulae for the radial variation of orbital frequency $\omega(r) = 2\pi/P(r)$ and the associated gradient shear $d\omega/dr$ (in units rad/yr/au).

2. After looking up the orbital distances d of each of the other seven planets besides Earth, compute each of their associated equilibrium temperatures.

3. Pluto has an eccentric orbit that ranges from a perihelion of 29.7 au to 49.3 au. Compute Pluto's equilibrium temperatures T at perihelion, aphelion, and its average distance.

Exercises

1. *Specific angular momentum and energy*

    a. What is the specific angular momentum (in both m^2/s and au^2/yr) your own body has due its motion with Earth around the Sun?

    b. If you flew radially from the Sun conserving this angular momentum, what would be your lateral speed v_{lat} (in cm/s) at a distance 1 pc from the Sun.

    c. What specific energy E (in $(m/s)^2$ = J/kg) do you need to reach this distance of 1 pc from Earth orbit at 1 au?

    d. What lateral speed v_{orb} (in m/s) is needed for maintaining a circular orbit at this 1 pc from the Sun, and how does this compare to angular momentum conserving lateral speed v_{lat} from part b?

    e. How much additional specific energy δE (again in $(m/s)^2$) is needed to bring you into circular orbit, and how does this compare to the E from part c?

2. *Transport of angular momentum from Keplerian shear*

    Consider material in a proto-stellar Keplerian disk in circular orbit at a radius r from a star with mass M.

    a. Derive formulae for the radial variation of orbital frequency $\omega(r) = 2\pi/P(r)$ and specific angular momentum $j = r^2\omega$.

    b. Now derive their associated gradients $d\omega/dr$ and dj/dr.

    c. For two layers with radii $r \pm \delta r/2$ separated by a small distance $\delta r < r$, which layer (upper or lower) has higher ω and which has higher j?

    d. Now imagine that diffusion causes equal mass parcels to exchange across a small length δr. Which layer, upper or lower, gained angular momentum, and which lost?

    e. Which layer will then tend to move inward and which outward?

3. *Albedo*

    a. Generalize the equilibrium temperature (Eq. equation (23.1)) to the case that a body has a nonzero albedo, $a > 0$, i.e. that it reflects a fraction a of incoming light, instead of absorbing it.

    b. Derive Earth's equilibrium temperature given its mean albedo $a \approx 0.3$.

    c. How does this compare to the Earth's mean temperature? What additional physics might explain the difference?

4. *Hydrostatic equilibrium in a gaseous disk*

    Consider the gaseous phase of a proto-planetary disk in Keplerian orbit around a proto-star with solar mass and luminosity?

    a. At a height z above a point in the disk plane that is a distance r from the star, what is the vertical (z) component of the stellar gravity, g_z?

    b. Applying this in the equation for hydrostatic equilibrium for a locally isothermal disk with sound speed c_s, derive the vertical variation of $P(r,z)$ in terms of its base value $P_0(r) \equiv P(r, z = 0)$.

    c. For the equilibrium radial variation of temperature, what is the radial variation of disk thickness $H(r)$?

    d. What is the associated radial variation of disk opening angle $\alpha(r) = H(r)/r$?

    e. Evaluate H and α at 1 au.

24 Water Planet Earth

24.1 Formation of the Moon by Giant Impact

The large number of moons of the giant planets like Jupiter and Saturn likely formed through angular momentum conservation during the gravitational contraction of their protoplanetary gas cloud, effectively making them each mini-planetary systems on a smaller scale. The size and mass ratios of these moons to their host planets are very small, much like the ratio of planets to the Sun.

In this respect, the Earth's Moon is quite distinct in being a comparable size ($\sim$1/4) to Earth. Samples from the Apollo missions show the Moon has an isotopic signature very similar to that of the Earth, indicating it likely formed from material in the Earth's crust and mantle. Because it also lacks much of the iron that makes up the Earth's core, this led to the theory that even well after (perhaps 1–10 Myr) the proto-Earth had formed and the heavy iron had settled into its core, there was a *giant impact* by a third, Mars-sized body – often dubbed "Theia." This impact ejected material from the Earth's mantle into orbit, which quickly cooled and condensed into the Moon (see Figure 24.1). Over billions of years, tidal coupling between the Earth and Moon transferred angular momentum from the Earth's rotation to the Moon's orbit, causing it to migrate from its initially close orbit to its present distance some 30 Earth diameters away.

While such a giant impact might seem rather unlikely and fine-tuned in the context of our present-day solar system, such events actually well represent the chaotic conditions that reigned during the final phase of planets clearing out competing large bodies that overlapped with their own orbit.

24.2 Water from Icy Asteroids

Of course, the energy and heating of such an impact likely had major consequences in also expelling much of the volatile material – such as water – that might have been retained from Earth's initial formation. But the extensive cratering on the Moon shows that even well after it formed, there were still ongoing extensive impacts from other bodies. This "Late Heavy Bombardment" is thought to have been triggered by ongoing gravitational interactions among the outer planets, leading to migration of Jupiter through the asteroid belt, and perhaps also a swapping of the orbital positions

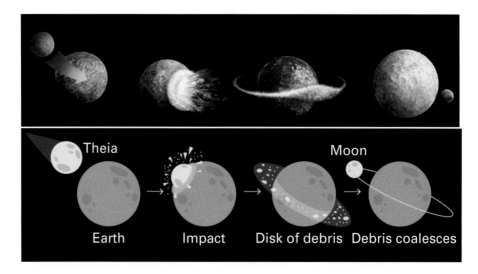

Figure 24.1 Illustrations of the giant-impact model for formation of the Earth–Moon system. Credit: Citronade, CC BY-SA 4.0, https://commons.wikimedia.org/w/index.php?curid=72720188.

of Uranus and Neptune. This sent the icy minor bodies hurtling toward the inner solar system, to impact the Moon, and, of course, also the Earth. In the vacuum on the Moon, heating by the Sun melted and evaporated the volatile water, which was then lost into space;[1] but on Earth, the ice from these ongoing impacts likely provided the source for much of the copious water that fills Earth's oceans today. Comparing the isotopic signature of ocean water with recent analyses of space-mission samples from comets and icy asteroids indicates that the latter are most consistent with providing most of Earth's water.

24.3   Our Magnetic Shield

Central to this retention of Earth's water is that fact that the Earth has also retained a dense atmosphere, which then keeps the water cycling among oceans, air, and land. The gases in the Earth's atmosphere can be traced to outgassing from volcanoes, which themselves are a consequence of the plate-tectonic collisions driven by escape of heat from radioactive decay in the Earth's mantle. Subduction of plates also recycles carbon back into the interior and so helps regulate greenhouse carbon in Earth's atmosphere.

By contrast, the lack of plate tectonics on Venus means its heat builds up, then is released every few hundred million years in violent, planet-wide epochs of volcanic eruption. The associated release of carbon dioxide, without the carbon recycling in

---

[1] Except, evidence from lunar orbiters now suggests, in perpetually shadowed craters near the lunar poles, where still-frozen ice might prove a crucial resource for future lunar exploration.

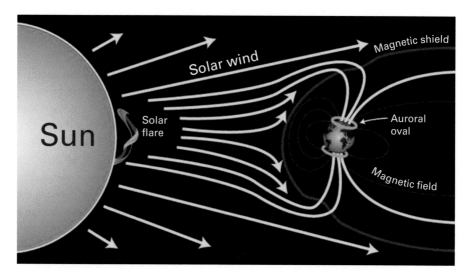

Figure 24.2 Illustration of how Earth's magnetic field shields it from direct impact by the solar wind. Credit: NASA.

plate subduction zones, has led to a runaway greenhouse effect, giving Venus the highest surface temperature in the solar system, some 700 K.

On the other side, the much smaller volume and mass of Mars implies its radioactive heating decayed away long ago, leaving now just a trace of extinct volcanoes. Moreover, the lack of a molten iron core meant there was no mechanism to produce the kind of global magnetic field that is generated by the convective dynamo in the Earth's core. Without such a global magnetic field, Mars is directly bombarded by high-speed protons from the solar wind, which over billions of years, has been steadily eroding Mars' initial atmosphere, so that it now has only about 1 percent the surface pressure of Earth's atmosphere.

As illustrated in Figure 24.2, the Earth's magnetic field effectively deflects these solar-wind protons, forming a magnetospheric shield to protect its atmosphere (and us) from their direct impact. Instead it just guides some fraction of solar particles to impact near the magnetic poles, where they harmlessly light up the upper atmosphere to form the beautiful dance of the northern and southern lights (the aurora).

24.4 Life from Oceans: Earth versus Icy Moons

Life on Earth originated in these oceans some 3+ billion years ago, first as single cells that gradually collected into evermore complex, multicellular forms. In the fullness of time, some grew a spine and crawled onto dry land, eventually leading even to large land animals, including humans like us. But even our bodies and cells still retain about 60 percent water by mass, reflecting their ancient origins in the oceans. Water is the essential solvent that transports the nutrients of life.

In addition to water, life fundamentally requires an energy source. For most present-day life, this can be traced to the energy from the Sun, captured via photosynthesis by green plants.

But an exception to this lies in deep-ocean vents, where the energy of upwelling magma seeds formation of complex, energy-rich compounds (e.g., hydrogen sulfide). These are then metabolized by giant worms and related organisms, stoking a complex ecosystem that is largely isolated and independent of life near the surface or on land.

While the icy moons of gas giants are too cold to have liquid water on their surfaces, the tidal flexing that arises from their eccentric orbits around their host planet can generate enough internal frictional heat to warm an extensive subsurface ocean. This is thought to occur in at least two such icy moons of gas giants. Europa orbiting Jupiter shows a complex, mottled surface of ice quite similar to ice flows seen in Earth's arctic oceans. And the relatively small satellite Enceladus orbiting Saturn shows cracks near its south pole, from which water geysers have been directly observed, and indeed directly sampled by passages through them by the Cassini spacecraft.

While Cassini did detect some of the simplest chemical building blocks of life, its instruments were not designed to detect more complex molecules, or life itself. Plans are currently being developed to send further spacecraft to both these icy moons, with the aim of directly detecting evidence for biochemistry or biological activity. If successful, this would, for the first time, extend our knowledge of life beyond our home planet Earth.

24.5   Questions and Exercises

Quick Questions

1. Estimate the mass of water in Earth's oceans, and how many comets had to hit the Earth to produce this water.

2. *Enceladus*
   Saturn's moon Enceladus has a radius of just $R = 237\,\mathrm{km}$, but exhibits water geysers.
   a. Assuming this moon's mean density is approximately that of water, determine the minimum speed (in m/s) for the ejected water to escape.
   b. How does this compare with the ejection speed from Old Faithful, which reaches a height up to 55 m?
   c. Discuss what factors (e.g., water temperature, gravity, air) may explain any difference.

Exercises

1. *Moon orbit*
   a. Given the Moon's period $P = 28\,\mathrm{days}$ and mean orbital distance $d_{\mathrm{m}} \approx 400\,000\,\mathrm{km}$, what is the Moon's orbital speed v_{m} (in km/s)?

b. Given the Moon's mass $M_m = 7.4 \times 10^{25}$ g, compute the Moon's associated orbital angular momentum $J = M_m v_m d_m$, in CGS units.

c. Approximating the Earth as a solid body with constant density, compute its moment of inertia $I = (3/5)M_e R_e^2$, given its mass $M_e \approx 6 \times 10^{27}$ g and radius $R_e \approx 6400$ km, again in CGS units.

d. Next compute the Earth's rotational angular momentum $J_e = I\omega$ (in CGS units), where its angular rotation frequency $\omega = 2\pi/(24$ hr$)$.

e. Suppose that the Moon first formed at a distance of just $d_o = 2R_e$, compute its orbital speed v_o.

f. Next, assuming the total angular momentum $J = J_e + J_m$ of the Earth–Moon system is conserved, estimate the Earth's rotation period when the Moon was at this distance.

g. Finally, what is the Earth's associated equatorial rotation speed, and what fraction is that of the speed needed to reach near-surface orbit?

2. *Hill radius of a planet*

Consider a planet of mass m that has a circular orbit of radius r around a star of mass $M \gg m$. The planet's Hill radius r_H is defined such that at this distance from the planet toward the star, the forces on an orbiting test mass will be in balance.

a. At such a distance r_H from the planet, and $r - r_H$ from the star, write out the combined acceleration g_{tot} from the star's gravity and the planet's gravity, as well as the centrifugal acceleration from orbiting the star with the same period as the planet.

b. Now set this $g_{tot} = 0$, and solve for r_H in terms of m, M, and r, under the approximations $m \ll M$ and $r_H \ll r$.

c. Compare this with the Roche radius computed in Exercise 2 of Chapter 7.

d. What is the Hill radius (in both R_e and au) for the Earth orbiting the Sun?

e. Finally, discuss the relevance of this and the associated "Hill sphere" in the formation of planets and their moon systems.

3. *Magnetic shield against the solar wind*

Interplanetary spacecraft at a Sun-distance $r = 1$ au measure typical solar wind proton density $n_p \approx 7$ cm^{-3} and speed $v_p \approx 450$ km/s.

a. For proton mass m_p, what is the associated mass density $\rho = m_p n_p$, in g/cm^3?

b. Next calculate the associated ram pressure $P_{ram} = \rho v_p^2$, in CGS units (dyne/cm^2).

c. The pressure of a magnetic field B, measured in gauss (G), is given in CGS units by $P_B = B^2/8\pi$. At its magnetic equator, the Earth has a field of about 0.3 G. What then is the ratio P_B/P_{ram}?

d. For dipole magnetic field, the field strength declines with distance d as $B \sim 1/d^3$. At what distance d from Earth (measured in R_e), is $P_B = P_{ram}$?

e. What is the minimum surface field strength B_{min} that Earth would have to have to keep this distance $d > R_e$?

25 Extra-Solar Planets

25.1 Direct-Imaging Method

Because they are much cooler than stars, the thermal emission from planets is mostly in the infrared. Their appearance at visible wavelengths comes instead by reflected light from their host star. This greatly complicates direct detection of extra-solar planets, since this reflected light is generally overwhelmed by the direct light from the star. Nowadays, there are some ($\sim$20) such direct imaging detections of exoplanets that appear far enough way from their host that it possible to block out the light from the star without also blocking the planet.

Figure 25.1 shows the example of a direct image, by one of the Keck Observatory's two 10 m telescopes, of four planets orbiting the star HR8799. The apparent shift in the positions of the planet images even allow one to infer their orbital periods, which range from 49 to 474 years.

But most exoplanet detections have been made via two other more-indirect techniques, known as the *radial-velocity* and *transit* methods, as illustrated by Figure 25.2.

Each of these three methods have analogs in the study of stellar binary systems, as outlined in Chapter 10. Direct imaging is similar to visual binary systems (Section 10.1); the radial-velocity method is similar to spectroscopic binaries (Section 10.2); and the transit method is analogous to eclipsing binaries (Section 10.3).

25.2 Radial-Velocity Method

As illustrated in the left panel of Figure 25.2, the radial-velocity method refers to the periodic movement of the host star due to the gravitational pull of the planet. The associated spatial "wobble" is not directly detectable but, as in spectroscopic binaries, its associated motion toward and away from the observer can be detected via very precise spectroscopic measurements of the systematic Doppler shift from multiple absorption lines in the star's spectrum.

Referring to Eq. (10.8) from our discussion of spectroscopic binary systems, let us identify object 1 with the planet of mass $M_1 = M_p$ and object 2 with the star that has

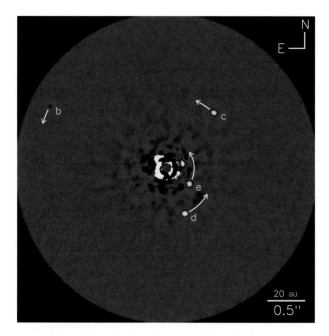

Figure 25.1 Direct image by the Keck Observatory of four exoplanets orbiting the star HR8799. The arrows indication magnitude and direction of their orbital motion from monitoring apparent shifts of their positions over more than a decade, yielding inferred orbital periods of 49, 100, 189, and 474 years for planets e, d, c, and b, respectively. For reference, the orbital period of the Sun's outermost official planet, Neptune, is 165 years. The planets are visible because most of the light from the star is blocked out by an occulting disk. The lower bar showing the extent of 20 au indicates the planets have orbital distance of several tens of astronomical units. Credit: NRC-HIA/C. Marois/W. M. Keck Observatory.

an orbital speed $V_2 = V_*$. Using the fact that $V_1/V_2 = M_2/M_1 = M_*/M_p$, we find for the star's speed

$$
V_* = \left(\frac{2\pi G M_{\text{tot}}}{P} \right)^{1/3} \frac{M_p}{M_{\text{tot}}}
$$

$$
\approx \left(\frac{2\pi G M_*}{P} \right)^{1/3} \frac{M_p}{M_*}
$$

$$
\approx 30 \, \text{km/s} \left(\frac{M_*/M_\odot}{P/\text{yr}} \right)^{1/3} \frac{M_p}{M_*}, \tag{25.1}
$$

where the second equality makes use of the fact that the planet's mass is negligible compared to the star, $M_{\text{tot}} = M_* + M_p \approx M_*$. This shows that the wobble speed is directly proportional to the planet's mass, but scales with the inverse cube root of the orbital period. The former favors planets with bigger mass, while the latter favors planets that orbit close to their parent star.

The very first detection of an exoplanet around another star was by this radial velocity method. It indicated what is now known as a *hot Jupiter*, i.e., a planet with a mass comparable to (actually even larger than) Jupiter, but orbiting at such a close distance that the stellar heating would make it quite hot. In the context of the prevailing idea (discussed in Section 23.4) that such gas giants should form only beyond the ice line, this detection was a real surprise. It led to a revised view that, while such hot Jupiters were indeed formed in the outer regions beyond the ice line, gravitational interaction with the proto-stellar disk and/or other giant planets out there led some to be flung into an *inward migration*, so that they finally ended up very close to their star.

Even if such complex interactions and migrations are rare, just the fact that they can occur at all can explain initial prominence of hot Jupiter detections, which according to Eq. (25.1) are, after all, observationally favored. This illustrates that, to assess the relative importance of such an observed phenomenon, it is important to be aware of the inherent observational biases that come with a given method or technique. It also means that it is helpful to identify other independent, and hopefully complementary, techniques.

For lower-mass planets orbiting with longer periods further from the star, the wobble velocity is smaller, and so can be hard to detect. For example, for the Earth orbiting the Sun, we have $P = 1$ yr and $M_e/M_\odot \approx 3 \times 10^{-6}$, giving $V_* \approx 9$ cm/s. With current technology, the smallest measurable speeds are about a factor 10 times higher; but because of the obvious interest in detecting Earth-size planets orbiting at a distance of 1 au around a Sun-like star, being able to detect wobble speeds near this Earth–Sun value is an ultimate, long-term goal.

25.3   Transit Method

Fortunately, there is another, quite-distinct way to detect exoplanets, known as the *transit method*. As illustrated in the right panel of Figure 25.2, this simply looks for the slight dimming of the star's apparent brightness whenever a planet "transits" in front of it. Instead of elaborate spectroscopic measurement of the slight Doppler shift, this merely requires precise photometric measurements of changes in the star's total apparent brightness. It is essentially analogous to eclipsing binaries discussed in Section 10.3 and illustrated in Figure 10.3, except that the lower-temperature star is now just replaced with a planet. Owing to its even-lower temperature, the planet emits mainly in the infrared, with little intrinsic emission in the visible. Thus the fractional drop in the star's observed optical apparent brightness is just set by the ratio of the projected areas of the planet versus the star, which scales with the square of the ratio of their radii,

$$\frac{\Delta F}{F} = \left(\frac{R_p}{R_*}\right)^2. \tag{25.2}$$

As illustrated in the right panel of Figure 25.2, the net change is thus greater for a smaller, red-dwarf (type M) star than for a larger, Sun-like (type G) star.

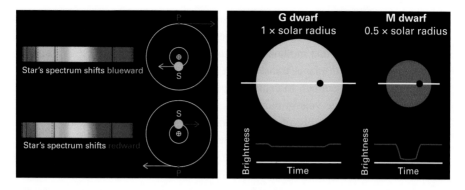

Figure 25.2 Left: Illustration of the radial-velocity method, showing how the orbit of a planet P causes a wobble of its star S; as the star alternatively moves toward or away from the observer (here to the left), this leads to small blueward or redward Doppler shifts of spectral lines in the star's spectrum. Note, the *dashed* lines here represent the original, *rest wavelength* positions of the spectral lines. Right: Illustration of the transit method, showing how the light from a star slightly dims when the dark planet goes in front of the star; the net effect is stronger for a cooler, smaller red-dwarf star (spectral type M) than for a solar-type G dwarf with higher temperature and larger radius.

The minimum size of planet that can be detected depends mainly on how precisely one can measure the stellar brightness. For ground-based telescopes, the main source of noise comes from distortions and variations from the Earth's atmosphere. The minimum planet-to-star size ratio scales with the square root of that noise. For example, a typical noise level of 1 percent allows one to detect a Jupiter-size planet, with $R_p/R_* \approx 0.1$. That is readily achieved even with ground-based telescopes.

But detecting smaller, rocky planets such as Earth, which have $R_p/R_* \approx 0.01$, requires a factor 1/100 lower noise, i.e., about 0.01 percent. This generally requires telescopes in space.

Another factor for the transit method is that it works only for planets whose orbital planes are near our line of sight. For a planet of radius r at a distance d from a star with radius R, the angle α that the line of sight makes to the planet's orbital plane must be within

$$\alpha_{max} = \arctan\left(\frac{R+r}{d-R}\right) \approx \frac{R}{d} = 0.0047\,\frac{R/R_\odot}{d/\text{au}} = 0.27\text{ degree }\frac{R/R_\odot}{d/\text{au}}. \quad (25.3)$$

The latter equalities show that detecting a transit of an Earth-size planet around a solar-type star, the alignment must be within an angle ± 0.27 degree. Over the full angle range from zero to 90 degrees, the associated probability of finding such Earth-like planet is only $P_e = 0.27/90 \approx 0.003$, or only 3 in every 1000.

Thus one generally has to monitor quite precisely a large number of stars to find those few that have this fortuitous alignment. A great breakthrough for this came from the Kepler satellite mission, named after the famous scientist who discovered the laws of planetary motion. Its prime mission was to monitor simultaneously the brightness of about 150 000 stars for several years, with a cadence of 30 minutes (and even every 2 minutes for a subsample). In addition to discovering planets, it also detected a wide range of phenomena associated with stellar brightness variations, such as starspots that modulate brightness with the star's rotation, or stellar pulsations. To discriminate planet transits from these other causes of variability, Kepler monitored stars long enough to capture repeat transits over several orbital periods, sometimes ranging up to year or more.

The probability of seeing a transit is highest for planets close to the star, which also tend to have the shorter, and thus more repeatable, periods. Thus most of the confirmed planets tend indeed to be from close orbits, and with large radii. But with extended monitoring over many years, it has become possible to identify a few planets with sizes near that of Earth, orbiting at distances up to an astronomical unit.

Unfortunately, degradation of Kepler's guiding gyroscopes eventually forced the so-called K2 phase of the mission to focus on different patches of sky, and finally to discontinue operations altogether. A follow-up mission called TESS (Transiting Exoplanet Survey Satellite) is systematically mapping the full sky, monitoring stellar photometric variability in the search for transiting planets.

25.4   The Exoplanet Census: 4000+ and Counting

As of this writing (Fall 2020), there have been 4000+ confirmed exoplanets, with several new ones added every day.[1] Figure 25.3 plots the overall population in terms of planet size (relative to Earth) and planet orbital period (in days). The legend for point style or color shows the discovery method, with the majority detected by the Kepler mission, based on the transit method. The radial-velocity method is quite successful at detecting both hot Jupiters and cold gas giants, but there are only a handful of planets found by direct imaging. Pulsar timing was actually the method for the very first detection, but it applies only in the rare case of a planet around a pulsar. A new method using gravitational microlensing shows promise for detecting relatively small planets far from their host star.

Figure 25.3 also outlines the various distinct classes of planets. While early discoveries were dominated by hot Jupiters and cold gas giants, transit surveys like the Kepler satellite have now identified numerous smaller planets. These range from the intermediate size ice giants, which when closer to their star become "ocean worlds," to rocky planets, which when very close can become "lava worlds," as the rocks are melted by intense heating from the star's radiation.

Note, in fact, the large number of planets detected with periods ranging from tens of days to even less than a day, implying they orbit much closer to their star than Mercury, with an orbital period of 88 days, is to our own Sun (0.4 au).

Such a more extensive survey with a variety of methods has given us a better understanding of observational biases. This allows one to compensate for the early

---

[1] A running compilation is given in
    https://exoplanetarchive.ipac.caltech.edu/docs/counts_detail.html.

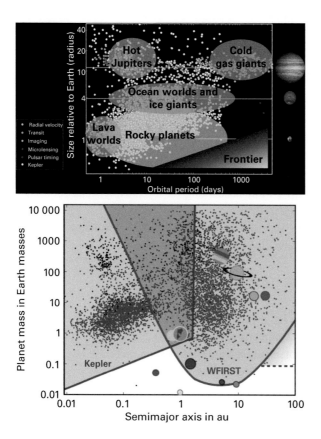

Figure 25.3 Top: Compilation of detected exoplanet populations, plotting planet size relative to Earth versus inferred orbital period. The color key gives the discovery method. Credits: NASA/Ames Research Center/Natalie Batalha/Wendy Stenzel. Bottom: Known planets discovered by the Kepler satellite (red) and all other methods (black), on a log–log plot of planet mass versus semimajor axis of orbit. The blue dots compare the predicted detections by the planned WFIRST satellite, using gravitational microlensing. Note that this method will favor detection of planets at much larger distances from their host star. The planet icons represent masses and orbital distances of planets in our solar system. Credit: NASA/WFIRST.

dominance of gas giants and hot Jupiters, which now are understood to be relatively rare. Indeed, the most numerous class are rocky planets somewhat larger than Earth, so-called super-Earths.

25.5 Search for Earth-Size Planets in the Habitable Zone

A key goal for ongoing exoplanet searches is to detect Earth-size planets in the "habitable zone," generally defined to be where liquid water could exist on the planet's surface. As noted in our discussion in Chapter 24 of our own Earth, the surface temperature can be affected by both the planet's reflection of visible light (e.g., from clouds), and the greenhouse trapping by the atmosphere of cooling radiation in the infrared. Since these are difficult to determine and quantify, a first approach is to assume that, as on Earth, these two effects roughly cancel, and so use just the simple blackbody equilibrium form derived in Section 23.5. For a star with surface temperature T_*, Eq. (23.1) can be generalized to

$$T(d) = 290 \, \text{K} \left(\frac{T_*}{T_\odot} \right) \sqrt{\frac{\text{au}}{d}} = 290 \, \text{K} \left(\frac{T_*}{T_\odot} \right) \left(\frac{M_\odot}{M_*} \right)^{1/6} \left(\frac{\text{yr}}{P} \right)^{1/3}, \tag{25.4}$$

where the latter equality uses Kepler's third law ($M \sim d^3/P^2$) to obtain a scaling with orbital period P, while accounting for a relatively weak additional dependence on stellar mass M_*.

For a lower-mass star with also a lower surface temperature, the period to have an equilibrium temperature the same as Earth is thus

$$\frac{P_e}{\text{yr}} = \left(\frac{T_*}{T_\odot} \right)^3 \left(\frac{M_\odot}{M_*} \right)^{1/2}. \tag{25.5}$$

For example, for a red-dwarf star with $M_*/M_\odot = T_*/T_\odot = 1/2$, we find $P_e = 65$ days, with a corresponding distance $d_e = 0.25$ au.

Such close-in, potentially habitable planets around cool, low-mass, red-dwarf stars are easier to detect, both directly and by the radial-velocity and transit methods; so there are already quite a few such candidates. However, because such cooler stars have deeper convection zones, they tend also to show quite extensive magnetic activity, with flares and coronal mass ejections. An ongoing area of study, known as "Living with a Star," seeks to examine how viable life could be on such close-in planets to active red dwarfs.

Another goal is to use subtle details of the observed spectrum from a star undergoing a transit to try to infer information on the planet's atmosphere, e.g., from absorption or the starlight by molecules in the planetary atmosphere that would not exist on the much hotter star. In particular, any signature of molecular oxygen would be viewed as an indicator for life, since this is normally very reactive and would be destroyed unless constantly being replenished by photosynthesis.

25.6   Questions and Exercises

Quick Questions

1. If we approximate the outer and inner limits of a habitable zone as ranging from where the equilibrium temperature is in the range $0 \, °\text{C} < T < 50 \, °\text{C}$, derive expressions, analogous to Eqs. (25.4) and (25.5), for the inner and outer values for the orbital period P_i and P_o, and for the orbital distances d_i and d_o.

2. A planet orbits a star that is twice as hot as the Sun at a distance of 4 au. Approximately what is the planet's equilibrium temperature?

3. a. If a Jupiter-size planet transits a solar-size star, by approximately what fraction is the star's light reduced?

   b. If an Earth-size planet transits a solar-size star, by approximately what fraction is the star's light reduced?

4. a. If a Jupiter-mass planet orbits a solar-mass star with a period of 1 yr, approximately what is the star's wobble speed?

   b. If an Earth-mass planet orbits a solar-mass star with a period of 1 yr, approximately what is the star's wobble speed?

Exercises

1. Using the orbital periods quoted in the caption of Figure 25.1, together with the projected separation distance from the central star, estimate the mass, in $M_\odot$, of the host star HR8799. Do this separately for each planet, and then compute the average mass and standard deviation.

2. The spectral type and color of HR8799 indicate it has an effective temperature $T_{\text{eff}} \approx 7400$ K. Assuming the planet separations shown in Figure 25.1 represent circular orbits seen face-on, compute the expected equilibrium surface temperatures T (in K) for each of the four planets labeled b, c, d, and e. Do you think life could exist on the surface of any of these planets or their moons?

3. *Radial-velocity detection of habitable planets*

   a. From Eq. (25.1) compute the amplitude of the Sun's wobble speed (in m/s) due to Jupiter's orbit. (You'll need to look up Jupiter's orbital period and its mass ratio to the Sun.)

   b. Ignoring Jupiter and other planets, similarly compute the amplitude of the Sun's wobble speed (in m/s) due to the Earth's orbit.

   c. The smallest wobble speed that can be currently measured in a star's spectrum is about 1 m/s. What does this imply about our ability to detect analogs of Jupiter and Earth around stars with mass comparable to our Sun?

   d. Next combine Eqs. (25.1) and (25.3) to derive an expression for the wobble speed of a star with mass M_* and temperature T_* due to a habitable zone planet of mass m_p and orbital period P_e.

   e. For a star with a mass and temperature that are both half that of the Sun, estimate the smallest mass planet M_p (in units of Earth's mass M_e) that could be detected in this star's habitable zone.

   f. How far (in au) would such a planet be from its star?

4. *Anthropic principle for our solar system*

   The anthropic principle explains any special conditions we find for our planet, our solar system, or even our universe as attributable to the special requirement that we as human beings must be able to evolve and survive in those conditions.

   a. Discuss this in the context of our home planet Earth having special conditions, including: distance from the Sun; prominence of water; existence of a large moon; an iron core that generates a magnetic field, etc. Could we have evolved and thrived on a planet that missed one or more of these features? Which ones seem essential, and which, if any, might not be?

b. Now do the same in discussing the properties of our solar system, containing: a star with the Sun's age, mass, luminosity, and temperature; inner rocky planets; a dominant gas giant such as Jupiter beyond the ice line. What do emerging results on exoplanet systems suggest for how typical or rare our solar system is compared with others?

c. Discuss finally what the above implies for how common large animals and beings like ourselves might be in the universe.

Part IV

Our Milky Way and Other Galaxies

26 Our Milky Way Galaxy

26.1 Disk, Halo, and Bulge Components of the Milky Way

The tendency for conservation of angular momentum of a gravitationally collapsing cloud to form a disk (see Section 22.5) is actually a quite general process that can occur on a wide range of scales: from planets, to proto-stellar cores, to even an entire proto-galaxy, with hundreds of billions times the mass of individual stars. This indeed provides the basic rationale for the disk in our own Milky Way (MW) Galaxy. We, along with our Sun, are today still embedded within the Milky Way's disk, orbiting about the galactic center, again because our bits of proto-galactic matter had too much angular momentum to fall further inward.

As we look up into a dark night sky, we can trace clearly the direction along this disk plane through the faint *milky glow* of thousands of distant, unresolved stars, from which we indeed get the name "Milky Way." Figure 26.1 gives a vivid illustration of the Milky Way through a panoramic image of the night sky seen from the exceptionally dark and clear site of the Paranal observatory, in the Atacama desert of Chile. Indeed, toward the horizon on the left one can also see two satellite galaxies of the Milky Way, known as the Large and Small Magellanic[1] Clouds (LMC and SMC).

As we look along this disk plane of the MW, the background/foreground superposition of many stars and giant molecular clouds (GMCs) makes it very difficult to discern the overall structure, the way we readily can from the face-on view of M51 in Figure 21.5. Moreover, the extinction from the extensive gas and dust means that visible images, like those in Figure 26.1 or in Figure 21.2, only penetrate a limited distance, typically $\sim$1 kpc, within the disk, which itself is only about 1000 ly, or just 0.3 kpc, in thickness.

Fortunately, infrared and radio images can penetrate much further, even to the other side of the galaxy, spanning the full 100 000 ly ($\sim$30 kpc) diameter of this disk. Thus with painstaking work applying various methods for determining the distance to the myriad of stars and GMCs detected, it has become possible to draw a quite complete map of the overall disk structure of our MW Galaxy, as given in Figure 26.2. This shows that, like M51, our Galaxy also has distinct spiral arms, along

---

[1] They are named "Magellanic" because they were first made broadly known to European civilization by Ferdinand Magellan, following his first-in-history circumnavigation of the Earth, with routes around southern continents showing the southern sky where these clouds are visible.

Figure 26.1 Panoramic photo of the Milky Way, taken from the European Southern Observatory's Very Large Telescope (VLT) facility in Paranal, Chile, located in the dry, and isolated, Atacama desert. The center shows silhouettes of four 8 m telescopes, along with a smaller 1.8 m telescope in the left background. Reflection of the laser light by the atmosphere is used in an *adaptive-optics* system by which secondary mirrors are deformed to correct for atmospheric seeing. On the far left, near the horizon, the two hazy patches are dwarf galaxies that are satellites of our own Milky Way, the Large and Small Magellanic Clouds. Though they appear to hang side by side, they are actually separated by about 15 kpc, and are removed from us by distances of 50 kpc and 60 kpc, respectively. Credit: A. Ghizzi Panizza/ESO.

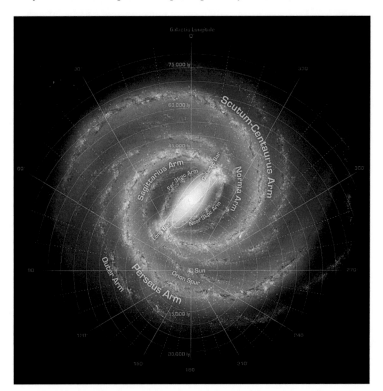

Figure 26.2 Map of the disk plane of our Milky Way Galaxy, based on IR and radio surveys. Our Sun lies along the Orion spur of the Sagittarius spiral arm, between the inner Scutum-Centaurus arm, and the outer Perseus arm. The galactic center (GC), toward the constellation Sagittarius, is defined to be at zero galactic longitude, with the Sun orbiting a distance $d \approx 8$ kpc from the GC, in the direction of longitude 90 degrees (i.e., to the left in the picture), toward the bright star Vega in the constellation Hercules. Credit: modification of work by NASA/JPL-Caltech/R. Hurt (SSC/Caltech).

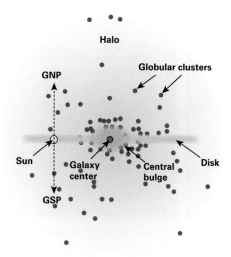

Figure 26.3 Edge-on schematic illustration of the 3D morphology of the MW Galaxy, showing the disk, halo, and bulge components, with globular clusters in halo, and the Sun in the disk, offset from the galactic center. The directions from the Sun away from the disk plane are dubbed the Galactic North and South Poles (GNP and GSP). Credit: RJHall at English Wikipedia.

which are concentrations of gas, dust, HII regions, GMCs, and active star formation. The map nicely illustrates the position of our Sun well away from the galactic center, and also serves to define the Sun-centered galactic longitude system used to chart the galactic disk.

Figure 26.3 illustrates schematically the overall 3D morphology of the MW, which in addition to the *disk*, has distinct *halo* and central *bulge* components.

The halo is roughly spherical, with a diameter comparable to that of the disk, about 30 kpc. It contains very little gas or dust, and thus without much source material for new star formation, its stars (dubbed "Population II") are very old. As discussed in Section 8.4, this can be seen from the H–R diagrams of the *globular clusters* (see Figure 8.1) that are common in the halo, which typically have main-sequence turnoff points below the luminosity of the Sun, implying ages $t > t_{ms, \odot} \approx 10$ Gyr. These old globular clusters contain of order $10^4 - 10^5$ stars, and are much more gravitationally bound and stable than the "galactic" or "open" clusters found in the disk.

Such open clusters are typically quite young, with main sequences that sometimes extend to masses of many tens of solar masses, implying ages less than their main-sequence lifetimes, i.e., less than a few tens of Myr. They are irregular in shape, and typically contain only 100 or so stars. They are so loosely bound that they tend to disperse within a few tens of Myr or less, evolving into unbound *OB associations*. Owing to tidal effects from the galaxy, along with the shear from its differential rotation, the stars eventually disperse and mix with other stars (called "Population I") in the disk. Figure 26.4 compares examples of a globular cluster

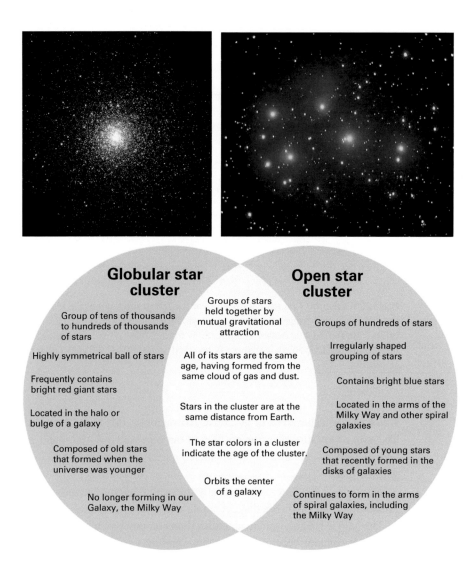

Figure 26.4 Top left: The globular cluster M80. Top right: The open cluster M45, also known as the Pleides or Seven Sisters. Bottom: A Venn Diagram comparing the different and common characteristics of globular versus open clusters.

and an open cluster, and gives a Venn diagram showing the common and distinct properties between the two types.

The central bulge contains a mixture of traits of both the disk and halo, with both types of clusters, and both populations of stars (I and II). Because of dust absorption from within the galactic disk, it does not appear in the visible to be much brighter than higher-galactic-longitude regions away from the galactic center; but if one corrects for this absorption, it dominates the overall galactic luminosity (as can be seen from M51, as shown in Figure 21.5).

26.2   Virial Mass for Clusters from Stellar Velocity Dispersion

By measuring the Doppler shifts of spectral lines from stars in an open cluster or a globular cluster, one can determine each star's radial velocity V_r. We can use this to define an average cluster radial velocity, $V_c \equiv \langle V_r \rangle$, as well as an *root mean square* (rms) *velocity dispersion* about this mean,

$$\sigma_v \equiv \sqrt{\langle (V_r - V_c)^2 \rangle}. \qquad (26.1)$$

The kinetic energy-per-unit-mass associated with this random component of radial velocity dispersion is $\sigma_v^2/2$. Assuming a similar dispersion in the two transverse directions that cannot be measured from a Doppler shift, the total associated kinetic energy from the three directions of motion is $K = (3/2)M_c\sigma_v^2$, where M_c is the total stellar mass. For a cluster of radius R, the associated gravitational binding energy scales as $U \approx -GM_c^2/R$. If the cluster is bound, then application of the usual virial condition $K = |U|/2$ allows one to obtain the cluster mass via

$$M_c = \frac{3\sigma_v^2 R}{G} = 6.9 \times 10^4 M_\odot \left(\frac{\sigma_v}{10\,\text{km/s}} \right)^2 \frac{R}{\text{pc}}. \qquad (26.2)$$

In practice, application of this method requires we obtain the cluster radius through the measured angular radius α and an independently known distance d, through the usual relation $R = \alpha d$.

26.3   Galactic Rotation Curve and Dark Matter

As illustrated in Figure 21.2, a primary diagnostic of atomic hydrogen in the disk plane of the Galaxy comes from its radio emission line[2] at a wavelength of $\lambda = 21$ cm. As we peer into the inner-disk regions of the Galaxy, i.e., along galactic longitudes in the range -90 degree $< \ell < 90$ degree, we find that this 21 cm line shows a distinct wavelength broadening $\Delta\lambda(\ell)$ that varies systematically with the longitude ℓ. Most of this broadening arises from cumulative Doppler shift along the line of sight from the motion associated with the orbit of distinct gas clouds about the galactic center; it thus provides a key diagnostic for determining the Galaxy's "rotation curve" as a function of galactic radius R.

Figure 26.5 illustrates the basic geometry and associated trigonometric formulae. Focusing for convenience on longitudes in the range $0 < \ell < 90$ degree, we find that the broadening actually takes the form of a redshift to maximum wavelength λ_{max}, which occurs when the line of sight along that longitude ℓ is *tangent* to some inner radius, $R = R_0 \sin \ell$, where R_0 (≈ 8 kpc) is the radius of our own galactic orbit along

---

[2] This results from a "hyperfine" transition in which the spin of the electron goes from being parallel to antiparallel to the spin of the proton. The energy difference is much smaller than for transitions between principal energy levels of the hydrogen atom, which are a few electron volts, and so have wavelengths of a few hundred nanometers, in the visible or UV spectral bands. See Question 2 at the end of the chapter.

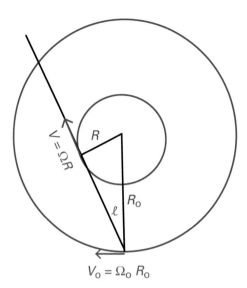

Figure 26.5 Sketch to show how measuring the change with galactic longitude ℓ of the maximum Doppler-shfted wavelength λ_{max} of the $\lambda_o \approx 21.1$ cm line from atomic hydrogen can be used to determine the galactic rotation rate $\Omega(R)$ as a function of radius R, given the known rotation speed $V_o = \Omega_o R_o \approx 220$ km/s at the radius R_o of the Sun's orbit. The results indicate that, inside the Sun's orbit ($R \leq R_o$), the rotation speed is nearly constant, with $V(R) \equiv R\Omega(R) \approx V_o$.

with the Sun. Thus by measuring λ_{max}, we can readily infer the *maximum* line-of-sight velocity away from us, $V_{rmax} = c(\lambda_{max}/\lambda - 1)$. Because our line of sight along ℓ is tangent to an orbit at this radius, the inferred maximum velocity just depends on the difference between the orbital velocity at R and the projection of our own orbital motion along this direction,

$$V_{rmax}(\ell) = V(R) - V_o \sin \ell = \Omega(R)R - \Omega_o R_o \sin \ell = (\Omega(R) - \Omega_o)R_o \sin \ell,$$
(26.3)

where Ω_o and $V_o = \Omega_o R_o$ are the angular and spatial velocity of the Sun's orbit at radius R_o. This can readily be solved to give the galactic rotation curve in terms of either the angular or spatial velocity

$$\Omega(R_o \sin \ell) = \frac{V_{rmax}(\ell)}{R_o \sin \ell} + \Omega_o \ ; \ \boxed{V(R_o \sin \ell) = V_{rmax}(\ell) + V_o \sin \ell.}$$
(26.4)

The Sun orbits the Galaxy at a radial distance $R_o \approx 8$ kpc from the galactic center, with a speed $V_o \approx 220$ km/s, implying then an orbital period $P_o \approx 220$ Myr.

Applications of this approach to analyzing observations of the 21 cm line of atomic hydrogen yield the rather surprising result that, within most of the region within the Sun's galactic orbit, $R < R_o$, the orbital speed is nearly the same as that of the Sun, i.e.,

$$V(R) \approx V_o \approx 220 \text{ km/s}; \ R < R_o,$$
(26.5)

which is known as a "flat" rotation curve.

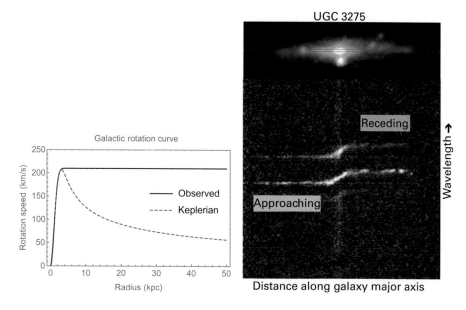

Figure 26.6 Left: Observationally inferred galactic rotation speed (in km/s, black line) versus radius R from the galactic center (in kpc), which shows a flat, or roughly constant, rotation speed for all radii $R > 5$ kpc. The red dashed curve compares the decline of speed as $V(R) \sim 1/\sqrt{R}$ that is expected from Keplerian motion with a mass that is as centrally concentrated as the stellar light. The difference implies there is a substantial "dark matter" contribution to the mass for $R > 5$ kpc. Right: The top panel shows a slit exposure of the galaxy UGC 3275 viewed with its disk edge-on to the observer line of sight. The lower panel shows the slit spectrum formed by plotting the wavelength spectrum of the star's light along the vertical, against distance along the major axis of the galaxy on the horizontal axis. The flat bright emission versus distance come from Doppler-shifted line emission lines that reflect the galactic rotation away from us on the right (longer wavelength) and toward us on the left (shorter wavelength). The flatness now shows quite directly that the rotation curve of this galaxy is also flat, as in our own Milky Way, again implying the presence of dark matter. Credit: Nicole Vogt, reproduced with permission.

Extension of this 21 cm method to longitudes 90 degree $< \ell < 270$ degree that point *outward* to larger galactic radii $R > R_0$ is complicated by the need now to have an independent estimate of the distance to an observed hydrogen cloud. But when this is done, the results indicate that the rotation curve remains nearly "flat," with *constant orbital speed*, out to the farthest measurable radii, $R \lesssim 15$ kpc. The left panel of Figure 26.6 compares this observed flat rotation curve for our Galaxy versus what would be expected from Kepler's law if the Galaxy's mass were as strongly centrally concentrated as its stellar luminosity. The right panel of Figure 26.6 shows that slit spectra across the edge-on galaxy UGC 3275 reveal similarly flat rotation curves in that galaxy.

The comparison with Kepler speed illustrates why these flat rotation curves came as a surprise. Since most of our Galaxy's *luminosity* comes from the central bulge within a radius $R_{bulge} \approx 1$ kpc, it seemed reasonable to presume that most of the Galaxy's *mass* would be likewise contained within this central bulge. But this would then require that galactic orbital speeds should follow the same radial scaling as derived for orbits around other central concentrations of mass, like the planets around the Sun. These follow the standard Keplerian scaling,

$$V_{\text{kep}}(R) = \sqrt{\frac{GM}{R}} \sim \frac{1}{\sqrt{R}}, \tag{26.6}$$

which would thus decline with the inverse-square-root of the radius.

Instead, the constant orbital speed V_0 of a flat rotation curve implies that the amount of mass within a given radius must *increase* in proportion to the radius,[3]

$$M(R) = \frac{V_0^2 R}{G} \sim R. \tag{26.7}$$

Since this extra gravitational mass extends to regions with very little luminosity, i.e., that are effectively very dark, it is known as *dark matter*. From studies extending up to scales well beyond our Galaxy, to clusters and superclusters of external galaxies, it is now thought that there is about *five* times more dark matter in the universe than the ordinary luminous matter that makes up stars, ISM gas and dust, planets, and indeed us. The origin and exact nature of this dark matter is not known, but it is thought to interact with other matter mainly just through gravity, and not through the electromagnetic and (strong) nuclear force that plays such a key role in the properties of ordinary "baryonic" matter.[4]

Nonetheless, as discussed below, this dark matter is now thought to be crucial to the formation of large-scale structure in the universe, and thus the associated galaxies that in turn provide the sites for formation of the stars, our Sun, and the planets such as our Earth. In short, without dark matter, we would not be here today to wonder about it!

26.4   Supermassive Black Hole at the Galactic Center

The center of our Galaxy is in the direction of the constellation Sagittarius, at a distance of about 8 kpc. Over this distance the absorption by gas and dust in the disk plane contribute to some $A_V \approx 25$ magnitudes of visual extinction. This corresponds to a reduction factor $F_{\text{obs}}/F_{\text{int}} \approx 10^{-A_V/2.5} = 10^{-10}$ in the visible flux, so almost completely obscuring this galactic center in the visible parts of the spectrum. But at longer wavelengths in the infrared and radio, for which the dust opacity is much lower, it becomes possible to see fully into the galactic center. Particularly noteworthy is Sagittarius A* (Sgr A), a region of very bright radio emission.

---

[3] For a spherical distribution of mass, the gravitational acceleration at any given radius R is just set by the mass $M(R)$ *within* that radius. For the visible mass in galactic disk the overall gravitational field is more complex. But, in practice, use of the simple spherical scaling form still provides a good approximation for mapping the overall distribution of *dark matter*, which is inferred to have a nearly spherical distribution in the galactic halo.

[4] Ordinary matter is often referred to as "baryonic" because most its mass comes from the protons and neutrons that are generally known as "baryons." Technically though, a small fraction of the mass of ordinary matter comes from electrons, which are actually classified as "leptons," not baryons.

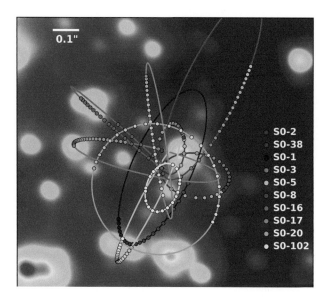

Figure 26.7 Subarcsec resolution of central arcsec of Sgr A, showing orbital tracks of individual stars about a common central point, now identified as the location of a supermassive black hole of mass $M_{bh} \approx 4 \times 10^6 M_{\odot}$. The individual dots show the annual positions of individual stars identified by the color-code legend over the 21-yr timespan 1995–2016. The star SO-2 has the shortest period, 16.17 yr, and so has been tracked over more than a full orbit. This image was created by Professor Andrea Ghez and her research team at UCLA from data sets obtained with the W. M. Keck Telescopes.

Infrared observations of the region around Sgr A shows a concentration of several hundred stars known as the "central parsec cluster." The blurring effects of the Earth's atmosphere normally limit spatial resolution to angular sizes of order an arcsec. But using specialized techniques – known as "speckle imaging" and "adaptive optics" – it has become possible over the past couple of decades to obtain IR images of individual stars within the central *arcsec* of the cluster, with angular resolution approaching ~ 0.1 arcsec. (The laser from the VLT in Figure 26.1 is an example of such adaptive optics in action.)

Monitoring[5] of the couple of dozen stars within this field since the mid-1990s has revealed them to have small but distinctive *proper motions*, following *curved orbital tracks* that all center around a common point just slightly offset from the Sgr A radio source. Figure 26.7 illustrates the tracks of 10 stars over the 21 yr period 1995–2016, with annual positions of the stars marked by dots, and individual stars identified by the color legend at the right. Using the known $d = 8$ kpc distance, the angular sizes of the orbital tracks can be translated into physical sizes for the semimajor axes a of the orbits. Extrapolating (or following) the motion over a full cycle allows one to infer the orbital periods P. Application of this and the semimajor axis into Kepler's third law

---

[5] Professor A. Ghez was one of the recipients of the 2020 Nobel Prize in Physics, for this work on monitoring stars orbiting the SMBH at the center of our Galaxy.

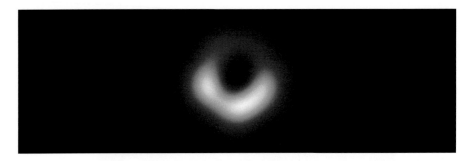

Figure 26.8 Direct silhouette image of the supermassive black hole at the center of the elliptical galaxy M87, as obtained by the *Event Horizon Telescope* (EHT). The EHT uses Very Long Baseline Interferometry (VLBI) to synthesize the 1.3 mm radio observations from eight radio telescopes spread around the globe, to approach the diffraction-limited angular resolution of a telescope with the aperture of the size of the entire Earth. M87 lies at a distance of about 17 Mpc, and the black hole is estimated to have a mass of $6.5 \times 10^9 \, M_\odot$, more than 1000 times the mass of the supermassive black hole at the center of our own MW Galaxy. Credit: Event Horizon Telescope Collaboration.

then gives an extremely large mass, $M_{bh} \approx 4 \times 10^6 \, M_\odot$ for the central attracting object, which is inferred to be a *supermassive black hole* (SMBH). Exercise 1 illustrates the process for this mass determination.

More information on these stars in the central pc can be found at the website for the UCLA Galactic Center Group:

`www.galacticcenter.astro.ucla.edu/`

Such SMBHs are actually inferred to be present at the centers of essentially *all* large galaxies. Figure 28.1 (in Chapter 28) gives an example of strong spectroscopic evidence for the active galaxy M84. But, as shown in Figure 26.8, there has recently been even a *direct imaging* of the silhouette of an even more massive SMBH at the center of the giant elliptical galaxy M87!

26.5   Questions and Exercises

Quick Questions

1. *Galactic year* At the distance $d \approx 8$ kpc of the galactic center, the Sun turns out to have an orbital speed $V_0 \approx 220$ km/s. How long is one "galactic year," i.e., the Sun's orbital period (in Myr) around the Galaxy?
2. *Energy of the hydrogen 21 cm transition* What is the energy E, in eV, of the hyperfine, spin-flip transition that gives rise to the 21 cm emission line of neutral hydrogen?
3. *Angular versus physical sizes at the galactic center*
   At the distance $d \approx 8$ kpc of the galactic center:
   a. What is the physical size s (in au) of an angle $\alpha = 1$ arcsec?
   b. What is the angular radius α_c (in arcsec) of the central parsec cluster?

Exercises

1. *Using Kepler's third law to infer mass of SMBH*

   To compute the mass of the SMBH at the galactic center, consider the star labeled SO-2 in Figure 26.7, which has recently completed a full, monitored orbit.

   a. Using the scale bar in the upper left showing the angular scale, estimate the angular extent (in arcsec) of SO-2's projected major axis.

   b. Using the known distance $d = 8$ kpc, what is the associated physical size s (in au) of the semimajor axis of SO-2's orbit?

   c. Next count the number of dots around the orbit to estimate the period P (in yr) of SO-2's orbit.

   d. Assuming we have a face-on view of SO-2's orbit, now use Kepler's third law to estimate the mass M_{bh} (in $M_\odot$) of the central black hole about which SO-2 is orbiting?

   e. Suppose our view is off by a modest inclination angle i from face-on. Does this increase, decrease, or have no effect on the mass estimate in part d? If it changes, by what factor does it change?

2. *Galactic orbital periods*

   Consider a model galaxy with $N = 210$ billion stars that have an average mass equal to the solar mass $M_\odot$ and that are *distributed uniformly* within a galactic disk of radius $R_{max} = 15$ kpc and thickness $Z = 0.3$ kpc.

   a. Derive a formula for the mass $M(R)$ (in $M_\odot$) inside of a given cylindrical radius R (in kpc) from the disk central axis. Quote results for both $R \leq R_{max}$ and $R \geq R_{max}$.

   b. Since $Z \ll R_{max}$, now ignore the disk thickness and assume all the mass at each radius R is concentrated at the disk midplane. For a test star in circular orbit within this midplane at any radius R from the center, derive a formula for the orbital period P (in Myr) versus radius R (in kpc). (Instead of looking up the gravitation constant G, etc., recall that 1 pc = 206 265 au, and use the fact the Earth's orbital period around the Sun is 1 yr.)

   c. Now derive an expression for orbital velocity $V(R)$ (in km/s) versus R (in kpc).

   d. For the Sun's orbital distance of $R_o \approx 8$ kpc, how do $P(R_o)$ and $V(R_o)$ compare with values for the actual sun?

3. *Galactic rotation from 21 cm radio observations*

   Suppose radio observations along the midplane of the Milky Way over a range in galactic longitude $\ell_{min} < \ell < 90$ degrees.

   a. Show that line emission from the atomic hydrogen (for which the rest wavelength $\lambda_o = 21.106$ cm) is shifted to a maximum wavelength that varies with longitude as

$$\lambda_{max}(\ell) \approx \lambda_o \left[1 + \frac{V_o}{c}(1 - \sin \ell)\right], \qquad (26.8)$$

   where $V_o = 220$ km/s is the Sun's orbital speed at a distance $R_o = 8$ kpc from the galactic center.

b. Use this to derive the galactic rotation speed $V(R)$ (in km/s) for radii R (in kpc) within the Sun's orbital radius R_o down to some minimum radius R_{min}.

c. Derive an expression for R_{min} in terms of R_o and ℓ_{min}.

d. From the results in part a, derive the mass $M(R)$ (in $M_\odot$) within any radius R (in kpc) over this same range from R_{min} to R_o.

4. *VLBI resolution of SMBH in M87*

Figure 26.8 shows the silhouette image of the SMBH at the center of the galaxy M87, obtained by the EHT telescope.

a. For the quoted inferred mass of the SMBH, compute its Schwarzschild radius R_{bh}, in km and au.

b. Noting that the last stable orbit around a black hole is 3 times R_{bh}, compute the *diameter* of this orbit, D_{orb}, again in km and au.

c. Next, for the quoted inferred distance $d = 17\,\mathrm{Mpc}$ to M87, compute the associated *angular* diameter of this orbit α_{orb}, in μarcsec.

d. Now, for EHT observations at wavelength $\lambda = 1.3\,\mathrm{mm}$, compute the diffraction-limit angular resolution, α_{EHT}, again in μarcsec. (Hint: Use Eq. (13.3). What is the effective telescope diameter for EHT's VLBI?)

e. Compare these values and discuss them in the context of the EHT image given in Figure 26.8. Does the silhouette really show the size of the SMBH?

5. *Virial mass of star cluster*

A cluster of $N = 20\,000$ stars within a cluster radius $R = 0.1\,\mathrm{pc}$ is observed in the hydrogen alpha (Hα) line with rest wavelength $\lambda_o = 656.3\,\mathrm{nm}$, and found to be in absorption from the spectra of the many individual stars. Averaged over the cluster, the line absorption is observed to extend from $\lambda_{min} = 656.6$ to $\lambda_{max} = 656.8$.

a. What is the mean radial speed $\langle V \rangle$ (in km/s) of the cluster relative to the observer? Is it moving toward or away?

b. What is the average velocity dispersion σ_v (in km/s) about this mean?

c. Use the answer in part b to compute the virial mass M_c (in $M_\odot$) for this cluster.

d. What is the average mass M_* (in $M_\odot$) of the individual stars in the cluster?

27 External Galaxies

27.1 Cepheid Variables as Standard Candles to External Galaxies

What we now know as external galaxies, like our Milky Way but far outside of it, were first identified by their signature spiral form. Known merely as "spiral nebulae," it was once thought they might be just stellar or cluster size regions like planetary nebulae, or the various other forms of diffuse nebulae seen in association with stars or star clusters.

The situation advanced considerably once telescopes became powerful enough to resolve individual stars within the great spiral nebula in Andromeda (see Figure 27.1). In the 1920s, using the 100 inch telescope on Mt. Wilson, Edwin Hubble was able to observe a particular kind of *pulsating* luminous giant star known as a *Cepheid variable*. Previous studies of Cepheid variables in our own Galaxy showed that they have the rather peculiar but very useful property that the period P of their pulsation – which can be readily measured – is related to their intrinsic luminosity L, or equivalently absolute magnitude M. Using a Cepheid with a measured period as a luminous *standard candle* with a known luminosity, Hubble's observation of the apparent brightness F (actually apparent magnitude m) of Cepheid stars within the Andromeda nebula led him to estimate its distance using the usual standard-candle formula,

$$d = \sqrt{\frac{L}{4\pi F}} = 10^{1+(m-M)/5} \text{ pc}. \tag{27.1}$$

The second equality uses the *distance modulus*, given by the difference between apparent and absolute magnitude of the Cepheid star, with the former observed and the latter inferred from the observed period and the Cepheid period–luminosity relation. The results indicated Andromeda was more than a million light years way! Since the Milky Way had already been inferred to have a diameter of only 100 000 ly, it was thus clear that Andromeda must lie well outside our galaxy, indeed with an angular size that implies it has a comparable physical size to the Milky Way itself.

Since this original application of Cepheid variables as standard candles, it has become clear that there are actually two distinct Cepheid classes: Types I and II, which apply respectively to Population I and II stars, with high and low metalicity. Figure 27.2 plots $\log L/L_\odot$ versus P (days) for Type I and Type II Cepheids, showing that the former are about a factor four more luminous at a given period. Hubble

Figure 27.1 The Andromeda galaxy (M31), the nearest large, external galaxy, at a distance of about 2.5 Mly from our Milky Way. The lower right-center also shows M33, the largest of several smaller satellite galaxies orbiting Andromeda, analogous to the Large and Small Magellanic Clouds (LMC and SMC) that orbit our own Milky Way. Credit: NASA.

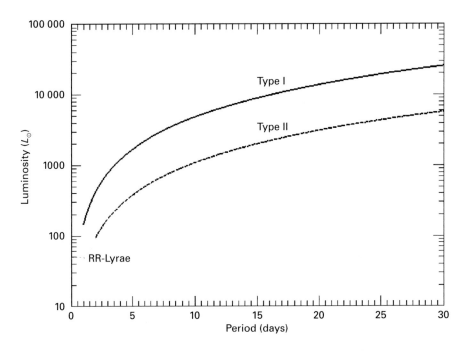

Figure 27.2 Period (in days) versus luminosity (in $L_\odot$, on a log scale) for Cepheid variables of Type I (high metalicity, of Population I, upper curve) and Type II (low metalicity, of Population II, lower curve).

incorrectly assumed that the Cepheids he initially used were of Type II, but they were actually of Type I. Accounting for the factor four higher luminosity within the observed apparent brightness implies that the Andromeda galaxy is actually twice as far as Hubble thought, i.e., some 2 Mly. Figure 27.1 shows a modern-day image of the full galaxy, with the spiral form that is thought to be quite similar to our own Milky Way.

Modern calibrations of classical (Type I) Cepheid variables give for the absolute visual magnitude M_v in terms of the period P,

$$M_v = -2.43 \log(P/10d) - 4.05 \tag{27.2}$$

For Type II Cepheids, the factor 4 lower brightness implies the magnitudes are 1.505 higher, meaning the -4.05 in Eq. (27.2) is replaced with -2.55.

27.2   Galactic Redshift and Hubble's Law for Expansion

As Hubble applied his Cepheid method to measuring distances to other spiral nebulae, a Mt. Wilson observatory night assistant named Milton Humason, a former mule driver without even a high-school diploma, became especially skilled at measuring their spectra from very faint images on photographic plates. In particular, he was able to measure the Doppler shift of known spectral lines, giving a direct measure of the galaxies' radial velocity V_r.

Quite surprisingly, Humason found that, with the exception of the relatively nearby galaxies such as Andromeda, all the more-distant galaxies showed only *redshifted* spectral lines, implying from the Doppler shift formula that they are all moving *away* from us, with $V_r > 0$.

Even more remarkably, when combined with Hubble's measurement of galactic distances, it led to what is now known as the *Hubble law*:[1] a *linear* proportionality between recession velocity V_r and and distance d,

$$V_r = H_0 d. \tag{27.3}$$

The proportionality constant, H_0, is known as the *Hubble constant*, which has units of *inverse time*. Figure 27.3 plots the original relations obtained by Hubble and Humason, with the slope of the red line fit to the data points giving H_0. Because of the incorrect assumption of the Cepheid type, along with a combination of other errors, the original value of nearly $H_0 \approx 500$ (km/s)/Mpc turns out to be a serious overestimate of the modern best value of $H_0 \approx 70$ (km/s)/Mpc.

The implications of this Hubble law are truly profound. In particular, if we simply assume that the velocity is constant, then dividing the distance by velocity gives the time since a distant galaxy was at zero distance from us,

---

[1] This is now sometimes referred to the Hubble–Lemaître law, because in 1927, two years before Hubble published his own article, the Belgian priest and astronomer Georges Lemaître published an article describing the law, but in French in a relatively obscure journal. It was thus overlooked until translated into English in 1931, two years after Hubble's paper.

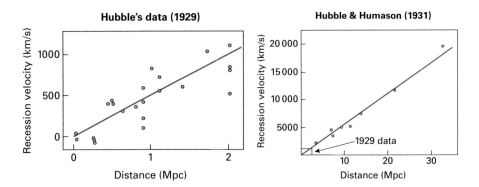

Figure 27.3 The original discovery forms of Hubble's law, showing a roughly linear proportionality between the recession velocity V_r of a galaxy and its distance. The slope of the red line fits through the data points gives a measure of the Hubble constant, $H_0 \approx 500\,(\text{km/s})/\text{Mpc}$. The modern best value is much smaller, $H_0 \approx 67\,(\text{km/s})/\text{Mpc}$.

$$t = \frac{d}{V_r} = \frac{1}{H_0} \equiv t_H \approx 10\,\text{Gyr}\,\frac{100\,(\text{km/s})/\text{Mpc}}{H_0}, \qquad (27.4)$$

where the third equality shows that this time, which is the same for all galaxies, is given by the inverse of the Hubble constant. This is known as the *Hubble time*, $t_H \equiv 1/H_0$, and as shown in the last relation of Eq. (27.4), a Hubble constant of $H_0 = 100\,(\text{km/s})/\text{Mpc}$ gives a Hubble time of approximately $t_H \approx 10\,\text{Gyr}$.

Modern observations of very distant galaxies show that the redshift,

$$z \equiv \frac{\lambda_{\text{obs}}}{\lambda} - 1 = \frac{V_r}{c}, \qquad (27.5)$$

can become quite large, with even some cases having $z > 1$. If taken literally in terms of the latter velocity Doppler shift formula in Eq. (27.5), this would seem to suggest that $V_r > c$, in apparent contradiction of special relativity.

But, as discussed in the cosmology sections in Part V, a more proper interpretation of this "cosmological redshift" is that it represents the stretching of the wavelength of light by the expansion of space itself! This can readily lead to redshifts $z > 1$. Einstein's limit really applies to how fast objects can travel relative to space, but that space itself can expand at a speed faster than light speed!

27.3   Tully–Fisher Relation

For more-distant galaxies, it becomes increasingly difficult to detect and resolve even giant stars like Cepheid variables as individual objects, limiting their utility in testing the Hubble law to larger distances and redshifts. For much larger distances, we need another, brighter "standard candle," like the white-dwarf supernovae (WD-SN) discussed in Section 31.1. But because the unpredictability of their appearance long

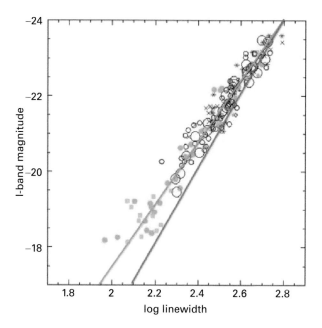

Figure 27.4 Empirical Tully–Fisher relation in the infrared, plotted as I-band absolute magnitude M_I versus logarithm of the total linewidth (in km/s), set by twice the rotational velocity, $2V_{rot}$. The blue line shows best fit line with slope −8.3, as given in Eq. (27.7), corresponding to velocity exponent 8.3/2.5 ≈ 3.3, slightly shallower than the slope 4 assumed in the simple $\log L_{gal}$ versus $\log V_{rot}$ scaling law of Eq. (27.6). The red line shows the slope −10 = −2.5 × 4 that would be implied in the I-band magnitude scaling by this simple form for the Tully–Fisher relation. The best-fit slope can differ for different wavebands. Credit: R. B. Tully.

limited the number of such WD-SN detections,[2] an important alternative method has been the so-called *Tully–Fisher* relation. Empirically, it was found that the infrared luminosity of a spiral galaxy, L_{gal}, scales with maximum rotation velocity V_{rot} inferred from the Doppler shift of spectral lines, with the approximate form

$$L_{gal} \propto V_{rot}^4. \tag{27.6}$$

The proportionality constant depends on the spectral band, but, as an example, in the near-infrared "I-band" (centered at 820 nm), the relation in terms of the absolute magnitude takes the numerical form,

$$M_I \approx -3.3 - 8.3 \log \left(\frac{V_{rot}}{\text{km/s}} \right). \tag{27.7}$$

Since magnitude $M \propto -2.5 \log L$, the slope of −8.3 here implies a velocity exponent 8.3/2.5 ≈ 3.3, somewhat shallower than the power 4 assumed in Eq. (27.6). Figure 27.4 shows actual I-band magnitude data versus $\log(V_{rot})$, compared with linear

---

[2] In the standard, formal notation, WD-SN are classified as "Type Ia," or "SN Ia."

relations with slope -8.3 (blue) and -10 (red), the latter corresponding to the simple case (Eq. (27.6)) with velocity exponent $4 = 10/2.5$.

To glean a possible physical rationale for this empirical Tully–Fisher relation, first note again that, by Kepler's law, the rotational velocity V_{rot} at an outer radius R scales with galactic mass M_{gal} as

$$V_{\text{rot}}^2 = \frac{GM_{\text{gal}}}{R}. \tag{27.8}$$

On the other hand, the galactic luminosity L_{gal} scales with the galaxy surface brightness I_{o} times the surface area πR^2 out to this outer radius,

$$L_{\text{gal}} \propto I_{\text{o}} \pi R^2. \tag{27.9}$$

Combining Eqs. (27.8) and (27.9) gives the scaling

$$L_{\text{gal}} \propto \frac{V_{\text{rot}}^4}{I_{\text{o}}(M_{\text{gal}}/L_{\text{gal}})^2}, \tag{27.10}$$

which recovers the simple scaling of Eq. (27.6) if we assume a constant value for the surface brightness times the square of the mass-to-light ratio, $I_{\text{o}}(M_{\text{gal}}/L_{\text{gal}})^2$. Models of galaxy formation have tried to explain why this should be true, but the results are tentative and not clearly established and accepted. Nonetheless, as a strictly *empirically* calibrated relation, this Tully–Fisher scaling provides a luminous standard candle to infer distances beyond the range accessible to the Cepheid method, and so allows a calibration of the Hubble law to moderately large distances and redshifts.

27.4   Spiral, Elliptical, and Irregular Galaxies

Galaxies can be generally classified by three distinct types of morphology: spiral, elliptical, and irregular.

Spiral galaxies are similar to our Milky Way, with distinct disk, halo, and bulge components. A spiral density wave in the disk forms the spiral arms that are the regions of active star formation out of the cold clouds of gas and dust. The tightness of the winding of the arms can vary, and sometimes emanate from a central "bar." The galaxy M51, also known as the "Whirlpool galaxy," shown in Figure 21.5, provides a good example of a typical spiral galaxy. As illustrated in Figure 26.2, our Milky Way Galaxy is thought to be a barred spiral.

Elliptical galaxies have a spheroidal shape, with different gradations of elongation from nearly spherical (E0) to highly extended (E5), as illustrated in Figure 27.5. Their stars are generally found to be Population II, and thus quite old with reduced metalicity. There appears to be a near absence of ISM gas or dust, and thus little or no new star formation. In these respects, elliptical galaxies are similar to globular clusters that orbit in the halo of our Milky Way, but much bigger and more massive. Their physical sizes can span a large range, from about 0.1 to 10 times size of the 100 000 ly diameter of our Milky Way, i.e., only 10^4 ly for "dwarf ellipticals"

Figure 27.5 Examples of three types of elliptical galaxies; figure taken from https://cas.sdss.org/dr6/en/proj/basic/galaxies/ellipticals.asp.

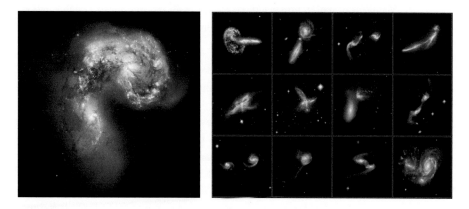

Figure 27.6 Left: The Antenna galaxy, which is actually two galaxies (NGC 4038 and 4039) undergoing a direct collision. Right: Gallery of other examples of interacting galaxies. Credit: NASA/ESA and the Hubble Heritage.

(with $M \sim 10^9 \, M_\odot$), to 10^6 ly for giant ellipticals (with $M \sim 10^{12} \, M_\odot$). At the center of a very large cluster of galaxies, there is often a giant "central dominant" (CD) elliptical galaxy that can have mass of $10^{12} \, M_\odot$ or more.

Irregular galaxies are just that. The overall structure is complex, though within subareas there can be spiral features. In many cases, it seems likely that the irregular form is because we are actually viewing two colliding galaxies, with then their mutual tidal interaction warping and disrupting whatever symmetric forms may have existed in the source galaxies. Figure 27.6 shows a mosaic of interacting galaxies (right), and an expanded view of the direct collision underway in the Antenna galaxies (left).

27.5   Role of Galaxy Collisions

For any collection of objects of size s separated by a mean distance d, the number density is $n \approx 1/d^3$ while the cross section is $\sigma \approx s^2$. The mean-free-path for collision is then

$$\ell = \frac{1}{n\sigma} \approx d \left(\frac{d}{s} \right)^2.$$ (27.11)

For individual stars, the distance/size ratio is enormous, of order $d/s \approx \text{pc}/R_\odot \sim 4 \times 10^7$, implying a mean-free-path $\ell \sim 10^{15}$ pc! Since this is more than 10^{10} larger than the $\sim$30 kpc size of a galaxy, we can conclude that individual field stars in galaxy should *never* collide.[3]

But for galaxies, this ratio of distance/size is much smaller, about a factor 20 for us to the Andromeda galaxy, and often just a factor of a few for galaxy clusters. Moreover, in the early universe, the average separation among galaxies not in the same cluster was smaller, with the factor reduction just set by the redshift, $1/(z+1)$. As such, while somewhat rare in the current-day universe, collisions can and do occur, and they were much more common in the early universe. Indeed, some models invoke a "bottom up" scenario in which larger galaxies form from the merger of smaller galaxies.

Animations from computer simulations of two colliding galaxies can be found on the web at:

        `www.youtube.com/user/galaxydynamics`.

The video dubbed "Spiral Galaxy" shows how spiral density waves can be induced by orbiting clumps of dark matter.

When galaxies do collide, their overall pattern of stars become strongly distorted by the mutual tidal interaction of the overall mass of the two galaxies; but the individual stars are too widely separate to collide, and so just pass by each other. In contrast, any gas clouds in the ISM of each galaxy do collide, with the resulting compression increasing the density of gas and dust, and thus often triggering a strong burst of new star formation. Such colliding systems are indeed often dubbed "starburst galaxies."

While distant galaxies show a redshift that implies they are moving away from us as part of the expansion of the universe, the mutual gravitational attraction between our Milky Way and the relatively nearby Andromeda galaxy is actually pulling them toward each other. Indeed, it now seems likely that Andromeda and the Milky Way will collide in about 3–4 Gyr. The "Future Sky" animation in the above link shows how the sky might appear from Earth during this collision.

---

[3] Some gravitational interaction can occur in the dense cores of compact globular clusters, but even there direct collision between stars is very unlikely.

27.6   Questions and Exercises

Quick Questions

1. Using Eqs. (27.1) and (27.2) compute the error factor f in distance if a Cepheid variable is assumed to be Type I but is actually Type II.
2. A Type I Cepheid variable is observed to have period $P = 20$ days and apparent visual magnitude $m_v = +10$. What is its distance (in pc)? Is this star in our galaxy?
3. A Type I Cepheid variable is observed to have period $P = 20$ days and apparent visual magnitude $m_v = +20$. What is its distance (in pc)? Is this star in our galaxy?

Exercises

1. *Angular size and surface brightness of Andromeda*
   The image in Figure 27.1 spans about 2.5 degrees across.
   a. Use this to estimate the angular diameter along Andromeda's long access. Compare this with the angular diameter of the full Moon.
   b. Assuming Andromeda has a similar physical size as our Milky Way, estimate its distance (in Mly).
   c. Now given its quoted distance of 2.54 Mly, estimate Andromeda's physical diameter in ly.
   d. Assuming Andromeda has a total luminosity of $10^{11} L_\odot$, compute its flux F in W/m^2 and associated apparent magnitude m.
   e. Finally, assuming that most of this flux originates in the core that's about 1/4 the full diameter, compute the surface brightness of this core, I_{core}, in W/m^2/sr. Compare the angular diameter and surface brightness of this Andromeda core to those for the Moon. (See Exercise 3.4 in Chapter 3).
2. *Application of the Tully–Fisher relation*
   A spiral galaxy with redshift $z = 0.23$ and apparent I-band magnitude $m_I = +17.6$ has an observed total spectral line width ratio, $\Delta\lambda/\lambda = 0.0013$. Compute the galaxy's:
   a. orbital velocity V_{rot};
   b. absolute I-band magnitude M_I;
   c. I-band luminosity L_I, in units of the I-band luminosity of the Sun, $L_{I,\odot}$ (for which the I-band absolute magnitude is $M_I \approx +4$);
   d. distance modulus $m_I - M_I$;
   e. distance d (in Mpc);
   f. recession velocity V_r from redshift z;
   g. associated Hubble constant $H_o = V_r/d$.
3. *Stellar collisions*
   Consider a model galaxy with $N = 210$ billion stars that each have the mass and radius of the Sun, and that are *distributed uniformly* within a galactic disk of radius $R_{max} = 15$ kpc and thickness $Z = 0.3$ kpc.

    a. Compute the disk stellar number density n (in pc^{-3}).

    b. Compute the mean-free-path for stellar collision.

    c. For a test star in a halo orbit perpendicular to the disk plane, compute the probability for it to collide with disk stars during any given disk passage.

    d. For a globular cluster of 10^5 stars with a galactic orbital period of 250 Myr, compute the average time for at least one of its stars to collide with a disk star.

28 Active Galactic Nuclei and Quasars

28.1 Basic Properties and Model

During the 1960s, sky surveys with radio telescopes discovered "QUAsi-StellAr Radio" sources, now known as "Quasars," or also Quasi-Stellar Objects (QSOs). In contrast to the extended radio emission sources seen from various regions of the Galaxy, these QSOs are, like stars, *point*-like sources without any readily discernible angular extent. They were soon identified with similarly point-like sources in the visible and other wavebands. But quite *unlike* stars, their spectral energy distribution does not even roughly match that of a blackbody of any temperature; instead it has an extended power-law form over a wide range of energies from the radio through the IR, visible, UV, and even extending to X-rays and gamma rays.

Nonetheless, this broad spectral distribution does still show patterns of absorption (and emission) lines that can be identified with known elements, but notably with a huge redshift z. For example, 3C273, one of the first and most famous QSOs, has $z = 0.158$, meaning that it is receding from us at a radial speed $v_r = 0.158c \approx 47\,000$ km/s. Taking a Hubble constant $H_o \approx 70$ (km/s)/Mpc, this puts its distance at $d \approx v_r/H_o \approx 677$ Mpc ≈ 2.1 Gly. With associated distance modulus $m - M = 5\log(d/10\,\mathrm{pc}) \approx +39$, together with its apparent magnitude $m = +15$, this implies an absolute magnitude $M \approx -24$, or luminosity $L \approx 5 \times 10^{11} L_\odot$! This far exceeds the luminosity of any star, and indeed even outshines the luminosity of a typical galaxy of $\sim 10^{11} L_\odot$.

Modern observations, e.g., with the Hubble Space Telescope, revealed that these quasars are commonly surrounded by a comparatively faint, diffuse stellar emission from a host galaxy. It is now realized that QSOs are indeed just one example of a class of "active galactic nuclei" (AGNs). In contrast to the extended galactic emission over a distance of a galactic diameter ~ 30 kpc, QSO/AGN emission is entirely point-like, emanating from the galactic nucleus. Indeed, since such QSO/AGNs often vary over timescales as short as a day, they must be very compact, no more than a light day in diameter, or $\lesssim 100$ au; this means they are roughly of order $\sim 10^8$ (~ 30 kpc/100 au) times smaller than their host galaxy.

The extreme luminosity from such a small volume is thought to be the result of matter accreting onto the supermassive black hole (SMBH) at the center of the QSO/AGN host galaxy. The SMBH in our Milky Way, and indeed in most galaxies in the nearby, current-day universe, are rather inactive, with relatively little ongoing accretion. But

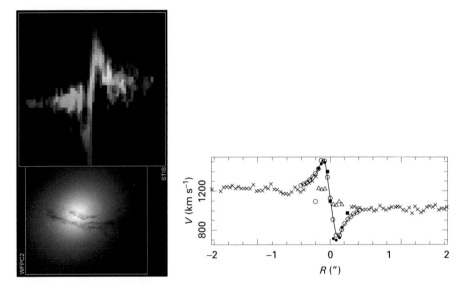

Figure 28.1 Slit spectrum (top left panel) made by HST's Space Telescope Imaging Spectrograph (STIS) of the active galaxy M84, as imaged in the lower left panel. The right figure plots the associated velocity variation along the galaxy's central axis. The high-velocity Doppler-shift near and across the galactic center implies the presence there of a supermassive black hole (SMBH). See Exercise 28.6. Credit: NASA/HST/STIS.

in the early universe, when the smaller intergalactic separation meant more-frequent galaxy collisions, the extreme disruption caused some stars to approach so close to the SMBH that they became tidally disrupted. The remnant stellar material typically still had too much angular momentum to fall directly onto the SMBH, and so instead fed an *accretion disk*. The viscous shear then transports angular momentum outward, allowing a steady, gradual accretion in which the gravitational energy released heats the disk and powers its emitted luminosity.

Figure 28.1 provides a good example of the direct evidence for such an SMBH at the center of the active galaxy M84. The upper left panel shows a color rendition of the Doppler shift seen from the slit spectrum shown in the lower left panel, which was placed to intersect the galaxy's center. The right figure plots the associated Doppler-shifted velocity versus position through the galaxy. Using the mean velocity to infer the redshift z and thus the distance, one can then use the velocity variation to infer the mass of the SMBH. Details are left as an exercise for the reader (see Exercise 7 at the end of this chapter).

Accretion down to the vicinity of a black hole can generate energy that is a substantial fraction of the rest mass energy of the accreting matter. (See Question 1 at the end of this chapter.) For an accretion rate $\dot{M}_{\mathrm{acc}}$ and conversion efficiency ϵ, the generated luminosity is

$$L_{\mathrm{acc}} = \epsilon \, \dot{M}_{\mathrm{acc}} c^2 = 1.4 \times 10^{12} L_\odot \frac{\epsilon}{0.1} \frac{\dot{M}_{\mathrm{acc}}}{M_\odot/\mathrm{yr}}. \tag{28.1}$$

The second equality shows the very enormous luminosity associated with accretion of 1 $M_\odot$/yr at a efficiency of $\epsilon = 0.1$ (the value for accretion to 5 Schwarzschild radii). It indeed readily equals or exceeds the luminosity inferred from the observed apparent magnitude and estimated distance of QSOs, including the example of 3C 273 mentioned above.

The SMBHs that power quasars are thought to be even more massive than those found in our and other nearby galaxies, of order a *billion* solar masses ($10^9 M_\odot$). But the associated Schwarzschild radii, $R_{bh} \sim 3 \times 10^9$ km ~ 20 au are still small enough to accommodate the day-timescale variation, even accounting for the fact that the emission region is likely to extend over $5-10 R_{bh}$.

28.2   Lyman-α Clouds

As this enormous luminosity from distant quasars propagates through the universe, it can sometimes pass through the relatively higher-density gas associated with galaxies or a galaxy cluster. Since the quasar spectral distribution extends well into the UV, the photons at wavelengths $\lambda = 121.57$ nm for the Lyman-α (Ly-α; $n = 1$ to $n = 2$) transition of neutral hydrogen, for which the opacity is very high, become strongly

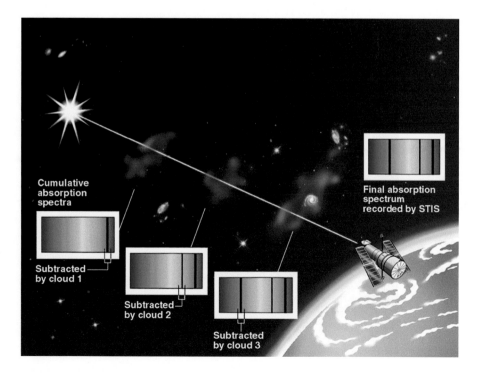

Figure 28.2 Illustration of Lyman-α clouds, in which hydrogen gas in galaxies at various redshifts absorb distant quasar light in the Lyman-α line from the $n = 1$ to $n = 2$ transition of hydrogen. Credit: NASA/STScI.

absorbed. But along this extended path length (Gly), the local Ly-α wavelength is Doppler shifted by the Hubble expansion, extending it to a longer wavelength that depends on the distance to the absorbing inter-galactic hydrogen cloud. As illustrated in Figure 28.2, this makes the observed quasar spectrum have a distinct number of absorption lines. Indeed, sometimes these are so dense that they are known as the Lyman-α "forest," with each "tree" of the forest corresponding to a distinct inter-galactic hydrogen cloud at a distance set by the cosmological (Hubble law) redshift of that observed absorption feature.

In essence, the huge luminosities and huge distances of quasars provide us a set of "flashlights" to probe the intergalactic hydrogen gas in the universe between us and the quasars.

28.3   Gravitational Lensing of Quasar Light by Foreground Galaxy Clusters

As this enormous luminosity from distant quasars propagates through the universe, it can also sometimes pass so close to a galaxy cluster that the gravity from the cluster's mass actually *bends* the rays of light, forming what is known as a "gravitational lens" (see Figure 28.3). This basic effect of gravitational bending of light was predicted by

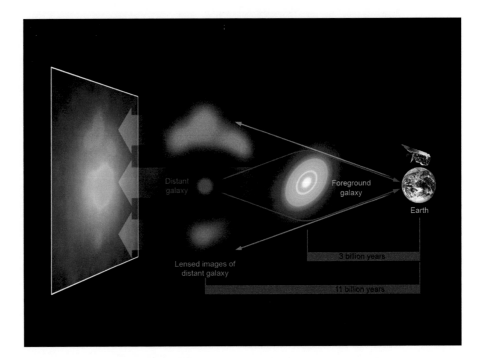

Figure 28.3 Diagram to illustrate "gravitational lensing," wherein the image of a distant galaxy or quasar is multiplied by the gravitational bending of light from passage near the very large mass of an intervening galaxy or galaxy cluster. Credit: NASA/JPL-Caltech.

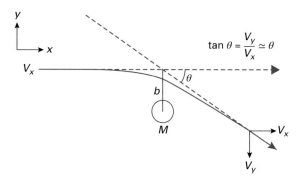

Figure 28.4 Diagram to illustrate the gravitational deflection of an object with initial speed V_x impacting within a distance b of a gravitational mass M.

Einstein's General Theory of Relativity, and was famously confirmed by expeditions to measure the associated shift in the position of stars as their light passed near the Sun during a solar eclipse. In the context of the passage of quasar light by a galaxy cluster, it can lead to multiple images, or even an "Einstein arc" or circle, if the quasar and galaxy's mass center are both closely aligned with the observer's light of sight.

Figure 28.4 illustrates the basic geometry and process. For an incoming velocity $v_x = c$ the deflection by mass M with impact parameter b results in vertical velocity deflection given by

$$v_y \approx \frac{GM}{b^2}t \approx \frac{GM}{b^2}\frac{b}{v_x} = \frac{GM}{bv_x}. \tag{28.2}$$

This resulting deflection angle is

$$\theta \approx \frac{v_y}{v_x} \approx \frac{GM}{bv_x^2} = \frac{GM}{bc^2}, \tag{28.3}$$

where the last expression gives the classical scaling for light with speed $v_x = c$. The full result from general relativity does have this scaling, but with a factor 4 correction,

$$\boxed{\theta = \frac{4GM}{bc^2}.} \tag{28.4}$$

28.4 Gravitational Redshift

At a distance R from a mass M, the gravitational binding energy per unit mass is $E_g = -GM/R$. An object launched from this distance with initial speed v_0 has specific kinetic energy $E_k = v_0^2/2 > -E_g$ and so total energy $E_{tot} = E_k + E_g$. It thus ends up with final kinetic energy $E_\infty = v_0^2/2 - GM/R = v_\infty^2/2$, and so final speed $v_\infty = \sqrt{v_0^2 - 2GM/R}$.

For light the speed always remains c, but whenever it propagates through a gravitational field, it too must conserve its total energy. A photon of wavelength λ

has radiative energy $E = hc/\lambda$, and so an associated mass $m = E/c^2 = h/(c\lambda)$. For an initial wavelength λ_o at radius R, this suggests an initial total energy $E_o = hc/\lambda_o - GMh/(Rc\lambda_o)$. To conserve this energy, the final wavelength far from the mass M must have a redshift given by

$$z + 1 = \frac{\lambda_\infty}{\lambda_o} = \frac{1}{1 - GM/(Rc^2)} = \frac{1}{1 - R_{bh}/(2R)} \approx 1 + \frac{R_{bh}}{2R} \; ; \; R \gg R_{bh}, \quad (28.5)$$

where $R_{bh} \equiv 2GM/c^2$ is the Schwarzschild radius. (See Section 20.4 and Eq. (20.3).) The above derivation is based on semiclassical physics arguments and so only applies for weak gravity cases, $R \gg R_{bh}$, which implies just a small redshift, $z \approx R_{bh}/2R \ll 1$.

From general relativity, the general formula for such *gravitational redshift* is given by

$$z = \frac{1}{\sqrt{1 - R_{bh}/R}} - 1. \quad (28.6)$$

For the nonrelativistic case $R \gg R_{bh}$, straightforward Taylor expansion leads to the simple, semiclassical form (Eq. (28.5)), with $z \approx R_{bh}/2R \ll 1$.

But as $R \to R_{bh}$, the redshift diverges, $z \to \infty$, enforcing the notion that light energy simply cannot escape from the Schwarzschild radius or below.

28.5   Apparent "Superluminal" Motion of Quasar Jets

The accretion that powers the tremendous luminosity of quasars can also drive relativistic jets from the polar axes perpendicular to the accretion disk. The jets emit in energies from the radio to gamma rays, but can be most finely resolved (to angular resolution less than a milliarcsecond (mas)!) spatially in the radio, using Very Long Baseline Interferometry (VLBI) from multiple radio telescopes spread across the Earth.

Indeed, such VLBI radio observations of such jets show they can be quite clumpy and variable on timescales of weeks to years. Quite remarkably, individual clumps in these quasar jets can sometimes show an apparent "superluminal" motion, meaning that, for the inferred quasar distance, the propagation of individual jet clumps away from the quasar can *appear* to be *faster* than the speed of light!

As illustrated in Figure 28.5, suppose a jet makes an angle θ with the direction to the observer on the right. If the motion is actually a fraction $\beta = v/c \lesssim 1$ that is near but below light speed c, then reductions in the light travel time give an *apparent* propagation speed fraction,

$$\beta_{app} \equiv \frac{v_{app}}{c} = \frac{vt \sin\theta}{ct - vt \cos\theta} = \frac{\beta \sin\theta}{1 - \beta \cos\theta}. \quad (28.7)$$

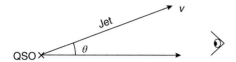

Figure 28.5 Illustration to show how quasar jet motion near the speed of light in a direction tilted toward the observer (on right) can lead to an apparent "superluminal" (faster than light) propagation speed away from the quasar.

As worked out in Question 4 below, the special case with $\beta = \cos\theta$ gives the maximum apparent speed, which in units of the speed of light is

$$\beta_{app}^{max} = \frac{\beta}{\sqrt{1-\beta^2}}. \tag{28.8}$$

From this it is clear that apparent superluminal propagation $\beta_{app}^{max} > 1$ is possible whenever the propagation speed $v = c\beta > c/\sqrt{2} = 0.707\,c$.

28.6   Questions and Exercises

Quick Questions

1. *Energy efficiency for accretion near a black hole*
   For accretion down to a radius that is a factor R_{acc}/R_{bh} times the Schwarzschild radius of black hole, compute the energy efficiency factor $\epsilon = E_g/mc^2$ for the gravitational energy gain E_b as a fraction of the rest mass energy mc^2 of the accreted mass. Confirm that $\epsilon = 0.1$ for $R_{acc}/R_{bh} = 5$.

2. *Variability constraint on black-hole mass*
   Suppose emission from a quasar comes from region that is 10 times the size (Schwarzschild radius) of its SMBH. What is the maximum SMBH mass (in $M_\odot$) consistent with a time variation of:
   a. 1 day
   b. 1 hr
   c. 1 week?

3. Show explicitly by Taylor expansion that the full general relativistic (Eq. (28.6)) for gravitational redshift recovers the semiclassical result (Eq. (28.5)) in the limit $R_{bh}/R \ll 1$.

4. For quasar superluminal motion, show explicitly that the maximum apparent fraction of the speed of light occurs when $\cos\theta = \beta$, and that this gives the β_{app}^{max} quoted in Eq. (28.8).

Exercises

1. *Lyman cloud speed and distance*
   Suppose a quasar shows absorption from a Lyman-α cloud at an observed wavelength $\lambda_{obs} = 183$ nm.

    a. What is the redshift z for this cloud?

    b. What is its inferred recession speed v_r (in km/s)?

    c. For a Hubble constant $H_0 = 70$ (km/s)/Mpc, what is its distance?

    d. How long ago (in Myr) did this light we observe leave this quasar?

2. *Gravitational lensing*

    Suppose a distant galaxy cluster with redshift $z = 0.2$ has two identical quasar images at equal angles $\theta = 10$ arcsec on each side of the cluster center.

    a. Assuming the current best value for Hubble constant, $H_0 \approx 70$ (km/s)/Mpc, what is the distance d (in Mpc) to the lensing galaxy?

    b. Use this distance d and the angle θ to estimate the closest distance b (in kpc) that the quasar's light passes to the center of the galaxy.

    c. Now use this b and the angle θ in Einstein's gravitational lensing formula to estimate the mass M (in $M_\odot$) of the lensing galaxy cluster.

3. *Gravitational redshift as alternative explanation of quasar redshift*

    Suppose we try to explain the redshift of 3C273 ($z = 0.158$) as a gravitational redshift, rather than being from cosmological expansion.

    a. Relative to the Schwarzschild radius R_{bh}, from what radius R_o is the radiation emitted?

    b. If the width of lines is 0.1 percent of their central wavelength, what is the range of radii (relative to R_{bh}) from which the radiation can be emitted?

    c. For a more physically reasonable assumption that any emission would come from at least radius range ± 10 percent around the central radius R_o, what would be the relative width $\Delta\lambda/\lambda_o$ of the observed emission line?

4. *Apparent superluminal motion in a quasar*

    Suppose VLBI radio monitoring shows that, over five years, a jet subcomponent of a quasar at a known distance $d = 1$ Gpc has moved away from the center by an angle $\Delta\alpha = 10^{-3}$ arcsec.

    a. What is the inferred apparent transverse speed of this component compared to the speed of light, $\beta_{app} = v_{app}/c$?

    b. What is the minimum actual fraction of the speed of light $\beta = v/c$ needed to give this apparent superluminal speed β_{app}?

    c. What is the associated angle θ (in degrees) between the component's motion and our line of sight?

5. *Quasar properties from redshift*

    A quasar of apparent magnitude $m = +15$ is observed to have its Hα line at a wavelength of 800.0 nm.

    a. What is the redshift z of this quasar?

    b. What is the speed V_r (in km/s) at which the quasar moving away from us?

    c. Assuming this redshift is due entirely to cosmological expansion with a value of Hubble constant $H_0 = 70$ (km/s)/Mpc, what is the quasar's distance ds (in Mpc) and distance modulus $m - M$?

    d. What is the quasar's absolute magnitude M and luminosity L (in $L_\odot$)?

6. *Luminosity from accretion onto a black hole*

   Consider a supermassive black hole of mass M_{bh} with an associated Schwarzschild radius $R_{bh} = 2GM_{bh}/c^2$.

   a. Assuming that a mass m dropped from far away down to a radius $R = 5R_{bh}$ is able to radiate all the energy $E = GM_{bh}m/R$ gained from gravity, compute the associated radiative energy conversion efficiency $\epsilon \equiv E/mc^2$. How does this depend on the black-hole mass M_{bh}? Explain.

   b. If mass accreted onto this black hole follows this efficiency, compute the luminosity L (in $L_\odot$) from an accretion rate of 1 $M_\odot$/yr.

   c. At what radius R_{bb} (in $R_\odot$) would a spherical blackbody of this luminosity have a temperature $T = 60\,000$ K?

   d. What black-hole mass M_{bh} (in $M_\odot$) would have $R_{bh} = R_{bb}/5$?

   e. If such an accreting black hole is observed to have a magnitude $m = +15$, what is its distance d (in Mpc)?

7. *Object at center of active galaxy M84*

   Figure 28.1 shows a Hubble Space Telescope slit spectrum taken of the nucleus of the active galaxy M84. Taking a standard value for the Hubble constant, $H_o = 70$ (km/s)/Mpc, use the information given in the figure to estimate the following (showing your calculations and explaining your reasoning):

   a. The distance to the galaxy.

   b. The conversion of angular scale (labeled in arcsec) along the horizontal axis into a physical distance, measured in both pc and au.

   c. The mass of the object at the very center (at $R = 0$ along the axis).

   d. How would your answers to a, b, and c differ if the actual value of Hubble constant were 10 percent lower or higher than the value assumed above?

29 Large-Scale Structure and Galaxy Formation and Evolution

29.1 Galaxy Clusters and Superclusters

Much as stars within galaxies tend to form within stellar clusters, the galaxies in the universe also tend to collect in groups, clusters, or even in a greater hierarchy of clusters of clusters, known as "*superclusters*." Our own Milky Way is part of a small cluster known as the "Local Group," which includes also the Andromeda galaxy, as well as several dozen smaller, "dwarf" galaxies. Along with roughly a hundred or so other groups, this makes up the "Local Super-cluster," with the highest concentration in the direction of the constellation Virgo. That concentration is also known as the Virgo (super)cluster, at a distance of about 20 Mpc, but its outer extent could be even be defined to include the Local Group, making it a possible center of the local supercluster. In any case, this is just one of millions of superclusters in the known universe.

Over the past couple of decades there have been several very large surveys that aim to measure the redshift of a large number (nowadays reaching many *millions*!) of galaxies along selected swaths of the sky. Over these large expanses of the universe, this measured redshift z gives, for a known Hubble constant H_o, a direct measure of the distance to the galaxy. For the simplest case of a constant expansion with Hubble constant H_o, Eq. (30.22) from Chapter 30 gives

$$d(z) = d_{\mathrm{H}} \left(1 - \frac{1}{1+z} \right) \tag{29.1}$$

$$\approx z\, d_{\mathrm{H}} ; \quad z \ll 1, \tag{29.2}$$

where $d_{\mathrm{H}} \equiv c/H_o \approx 14\,\mathrm{Gly}$ defines the "Hubble distance," i.e., the distance that light travels over a Hubble time $t_{\mathrm{H}} = 1/H_o \approx 14\,\mathrm{Gyr}$. The latter equality shows that, for small redshift, this just recovers the standard result that distance is just linearly proportional to redshift.

With the readily measured two-dimensional (galactic longitude and latitude) positions on the sky, a 2D survey along a swath on the sky can be combined with the redshift distance to form a three-dimensional (3D) picture of the universe through that swath. The upper-left panels of Figure 29.1 (blue color) show a slice of this 3D picture containing one dimension of galactic position plus the distance, arranged along the radius from our own observer's position at the origin. The result shows a remarkable

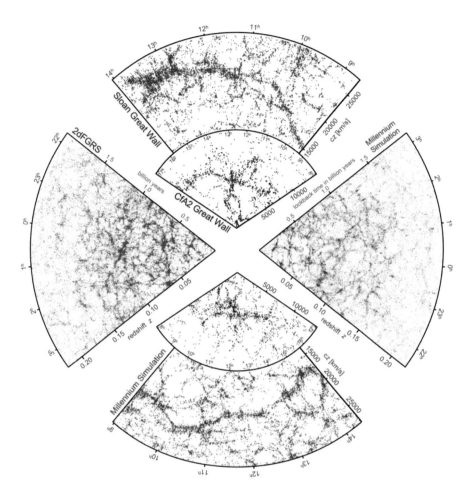

Figure 29.1 Comparison between observational surveys (blue, top and left) versus computer simulations (red, bottom and right) of the large-scale structure of the local universe. The observational surveys measure position and redshifts of millions of galaxies along an extended, narrow arc of the sky, using the redshift z to estimate galactic distance. The simulations assume initial seed perturbations set to correspond to those inferred from Cosmic Microwave Background (CMB) fluctuations, plus cold dark matter to enhance gravitational attraction, and then compute gravitational contraction of structure starting from nearly uniform early universe at redshift $z > 25$ to the present, highly structured, local universe with redshift $z < 0.25$. Credit: Volker et al. 2006, *Nature*, 440, 1137.

"cosmic web" in the overall large-scale structure (LSS) of the universe. This has a concentration of galaxies along extended, thin "walls," surrounding huge voids with few or no galaxies in the huge volume between the walls. But there are particularly high concentrations at the intersections of the walls. Indeed, most previously identified superclusters can be associated with one of these wall intersections.

The lower-right panels of Figure 29.1 (red color) show the results of very large simulations for the formation of the structure from the gravitational attraction by matter. For one such simulation, Figure 29.3 shows a sequence of volume renderings at different stages of the formation of structure, identified by the redshift z associated with each epoch. As shown in the upper left panel for the earliest phase of the simulations at a redshift $z = 27.30$, one also requires a small initial seed of density fluctuations, which are then amplified by the gravitational attraction. As discussed in Chapter 32, it is now thought that this initial seed of small-amplitude variations in density is provided by quantum fluctuations in the very early phases of the Big Bang itself!

29.2   Lensing of Colliding Galaxy Clusters Confirms Dark Matter

Much like the individual galaxy collisions shown in Figure 27.6, the random motions of whole *galaxy clusters* in the early universe means they can also collide. Figure 29.2 shows the example of such a collision in the distant ($z = 0.586$) supercluster dubbed MACSJ0025.4-1222.

The clusters are inferred to be colliding toward each other along an axis that appears roughly horizontal in the plane of the picture. The red halo in the center indicates X-ray emission from hot, intergalactic gas that has been shock-heated and compressed by the collision. The broader lateral extent of individual galaxies observed with the Hubble telescope show that, because of the space between them and their constituent stars, these have essentially just passed through each other, without any compression by the collision.

The blue halo indicates the cluster mass distribution inferred from mapping the gravitational lensing of a background source. It is identified with the dark matter that binds the galaxy clusters, and the fact that it too is extended beyond the central halo of X-rays indicates that, unlike the hot intergalactic gas, it does not self-interact through the forces that are key in gas collisions.

Such examples of noninteraction of lens-inferred dark matter are found in other collisions of galaxy clusters, including the initial discovery in 2005 of such a case in the somewhat closer (3.7 Gly) and more-well-known Bullet cluster of galaxies.

Overall, this provides a strong, independent confirmation for the reality of dark matter.

29.3   Dark Matter: Hot versus Cold, WIMPs versus MACHOs

A key result of the simulations shown in Figure 29.3 is that achieving an LSS that has the same statistical form as the observed structure requires inclusion of a significant component of *cold dark matter* (CDM), with a total mass that is factor several (about five) times the mass of ordinary matter that makes up the planets, stars, and galaxies that produce the various spectral bands of electromagnetic radiation we directly observe. "Cold" here means that the matter is nonrelativistic, so that its gravitational contribution comes from its rest mass, and not from any relativistic enhancement in

Figure 29.2 Collision of two galaxy clusters (dubbed MACSJ0025.4-1222) induce X-ray emission (red) from hot, shock-compressed intergalactic gas, as observed by NASA's Chandra X-ray Observatory. The individual galaxies, imaged by the Hubble Space Telescope, pass right through each other, and so remain extended along the (horizontal) collision axis. The blue halo shows the dark matter distribution inferred from gravitational lensing of background sources. The fact that it too is extended along the collision axis indicates that, unlike the hot gas, it does not self-interact like ordinary matter. This represents a key, independent line of evidence for dark matter. Credit: X-ray(NASA/CXC/Stanford/S.Allen); Optical/Lensing(NASA/STScI/UC Santa Barbara/M.Bradac).

its energy. Only CDM seems able to form the deep gravitational wells from mutual attraction to frame the observed wall plus void network of galaxies observed for large-scale structure. Hot dark matter tends to remain too distributed.

There are two candidates for CDM, dubbed by the somewhat whimsical terms "WIMPs" – for Weakly Interacting Massive Particles – and "MACHOs" – for MAssive Compact Halo Objects. The latter refer to a conjectured large population of low-mass objects – perhaps roving Jupiter-size bodies that are too cold to emit much radiation – thought to occupy the core and halo of galaxies. To explain the flat

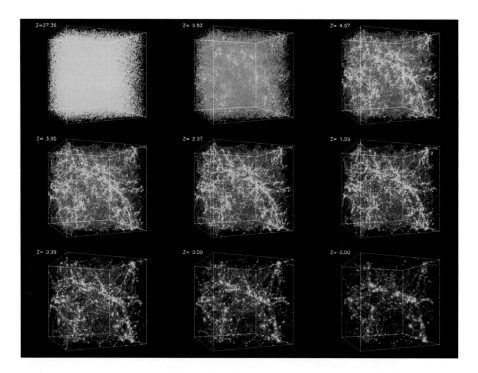

Figure 29.3 Computer simulations of evolution of large-scale structure in a universe, beginning with the nearly smooth, early universe at redshift $z = 27.36$ (upper left) to the extensive structure in the current local universe at $z = 0$ (lower right). The so-called ΛCDM model assumes cold dark matter and dark energy. Credit: Center for Cosmological Physics, University of Chicago. Simulations were performed at the National Center for Supercomputer Applications by Andrey Kravtsov (University of Chicago) and Anatoly Klypin (New Mexico State University). Visualizations by Andrey Kravtsov.

rotation curves of galaxies, the number density of such MACHOs would have to be so large that as they, randomly pass in front of stars, they should induce a gravitational "microlensing" event that should be observable from monitoring of the star's light. Extensive monitoring surveys have indeed detected such microlensing events, but at a rate that is well below what would be needed for MACHOs to be a significant component of dark-matter mass.

There is thus now a general consensus that CDM most likely consists of some kind of WIMP. "Weakly interacting" in this context means they are not subject to either the *strong nuclear* force, which binds the nucleus of atoms, or *electromagnetic* forces, which bind electrons to atoms, and are responsible for producing light and all other forms of electromagnetic radiation. The inability to produce light is indeed what makes WIMPs a candidate for dark matter. Like neutrinos, they might be subject to the *weak* nuclear force, but otherwise they interact with ordinary matter only via gravity. Moreover, while neutrinos have a rest mass only only a few electron volts, a hypothetical WIMP could have a much larger mass, perhaps many hundreds times the GeV mass of protons and neutrons. There are several projects underway to detect

WIMPs, through experiments done deep underground, which shields against the flux of cosmic rays that would otherwise contaminate detections of the very few weak interactions by WIMPs.

29.4   Galaxy Evolution over Cosmic Time

An area of particularly active current research regards the origin, formation and evolution of galaxies over cosmic time. Figure 29.4 schematically illustrates the overall evolution of the universe from our current modern era populated by "normal" galaxies, back in time. This follows increased redshift and distance, to the most distant early galaxies observed by the Hubble Space Telescope (HST), and beyond to the first galaxies and even to the very first stars. As detailed in Section 33.2, about 400 000 years after the Big Bang, there was a recombination of the initially ionized plasma of protons and electrons, after which the universe became transparent to the CMB.

The time from this recombination to an age of about 100 Myr, when the first stars were born, is known as the "dark ages." There are ambitious efforts underway to

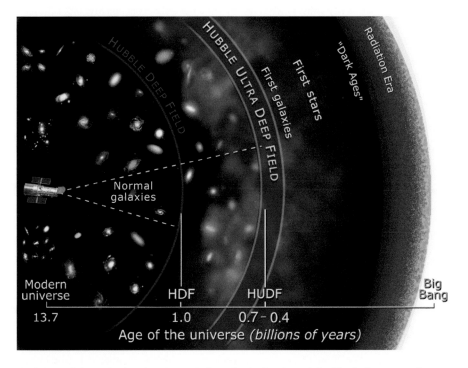

Figure 29.4 Schematic showing the evolution of galaxies through lookback time, extending to the radiation era of the Big Bang. The "dark ages" refer to the time when the universe consisted of neutral hydrogen gas, before the UV radiation from the first stars reionized the universe. As discussed in the text and illustrated in Figure 29.5, the Hubble Ultra Deep Field reaches back nearly to the time of formation of the first galaxies. Credit: NASA and A. Feild (STScI).

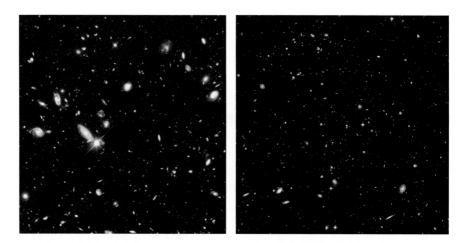

Figure 29.5 The Hubble Deep Field (HDF, 1995; left) and Hubble Ultra-Deep Field (HUDF, 2005; right), showing a myriad of faint distant galaxies in otherwise dark patches of sky with few or no foreground stars. The HDF includes more than a thousand galaxies within a 1 arcmin square patch of sky in Ursa Major. The HUDF is from the combined exposure time of 11.3 days of a 3 arcmin square patch of sky (about 0.1 the angular diameter of the Moon) in the constellation Fornax. Analysis shows the HUDF contains more than 10 000 galaxies, ranging up to 13.3 Gly away, to when the universe was only about 400 Myr old. Blue and green correspond to colors that can be seen by the human eye, originating from hot, young, blue stars as well as from cooler stars in the disks of galaxies. Red represents near-infrared light, which can originate from dust-enshrouded galaxies. Credit: NASA, ESA, and S. Beckwith (STScI) and the HUDF Team.

detect these first stars, or at least indirect signatures of their formation; for example by looking at what redshift the rest-wavelength 21 cm radio emission from neutral hydrogen gas is truncated, one can estimate the era of reionization by the UV light from the first generation of what are expected to be massive, luminous, hot stars.

There are also efforts to detect directly the most distant and thus very faint first galaxies. So far the deepest detections have been from the Hubble Ultra Deep Field, shown in Figure 29.5. Because of its very dark background sky, without the atmospheric scattered light of ground-based telescopes, the HST was able to combine many ($\sim$800) long time ($\sim$ half-hour) exposures, for a total exposure time of $\sim$11.3 days, to detect very faint, very distant galaxies. The patch of sky is about 3×3 arcmin, about 1/10 the angular diameter of the Moon, chosen because it contains very few foreground stars from within our own Milky Way galaxy.

But analysis shows this small patch contains more than 10 000 galaxies! The nearest ones, less than 1 Gly away, appear quite similar to other galaxies in our current-day universe, having mostly forms that range from spiral to elliptical. But the most-distant ones, ranging beyond 13.3 Gly away, have a variety of irregular forms, shapes, and sizes. These oddball galaxies chronicle a period when the universe was younger and more chaotic, when order and structure were just beginning to emerge.

Building on this pioneering study by Hubble, ground-based surveys have focused on cataloging how galaxy morphology changes with redshift, and thus over cosmic time. The goal is to understand how dark matter and the ordinary, baryonic matter it attracts evolves with time, first forming dark-matter halos, and eventually galaxies and galaxy clusters.

29.5   Questions and Exercises

Quick Questions

1. For the presently favored Hubble constant $H_o = 70\,(\text{km/s})/\text{Mpc}$, compute the Hubble time t_H (in Gyr) and Hubble distance d_H (in Gly).

2. At what redshift z is the approximate distance formula (Eq. (29.2)) 10 percent different from the full result (Eq. (29.1))? Is the approximation too small or too big?

3. For a Hubble constant $H_o = 75\,(\text{km/s})/\text{Mpc}$, what is the distance (in Mpc) of a quasar with redshift $z = 1$? How long ago did light leave this galaxy?

4. The most distant quasar known so far has a redshift $z = 7.54$. Assuming a constantly expanding universe with $H_o = 70\,(\text{km/s})/\text{Mpc}$, about what was the age t (in Gyr) of the universe when the light we see from this quasar was emitted?

Exercises

1. *Collision of galaxy supercluster*

   Figure 29.2 shows a composite of the colliding galaxy supercluster MACSJ0025.4-1222.

   a. For its observed redshift $z = 0.586$, what is its distance d in Gpc, assuming the standard value for $H_o = 70\,(\text{km/s})/\text{Mpc}$?

   b. How long ago t (in Gyr) did the light we see leave this supercluster?

   c. Given that the full frame shown is about 3×3 arcmin, what is the associated solid angle Ω (in sr) and the fraction of the full sky $\Omega/4\pi$?

   d. What is the associated physical size s (in Mpc and Mly) range across the image, and approximately how far apart Δs are the two blue lensing clouds of dark matter?

   e. Approximately how fast V (in km/s) would the colliding hydrogen gas clouds have to be moving to produce the X-rays observed with energy $E \approx 5\,\text{keV}$? (Hint: Set $m_H V^2/2 = E$.)

   f. At this speed V, how long a time Δt (in Myr) did it take the dark matter to cross this separation Δs?

2. *Number of galaxies in the universe*

   The HUDF in Figure 29.5 contains $\sim 10\,000$ galaxies within a patch of sky of 3×3 square arcmin.

   a. What is the solid angle Ω_{HUDF} of this patch (in sr)?

   b. What fraction f of the entire sky does this represent?

c. Use this value to estimate the total number of galaxies (N_{gal}) that would be observed from similar patches over the full sky.

d. Assuming the HUDF detects all galaxies out to a distance of ~ 10 Gly, estimate the average number density (n_{gal}) of galaxies in the universe (in Mly^{-3}).

e. What does this imply for the average separation distance d_{gal} (in Mly) between galaxies? How does this compare to the distance d_{And} between the Milky Way and Andromeda?

f. Discuss potential reasons for any differences between d_{And} and d_{gal}.

Part V

Cosmology

30 Newtonian Dynamical Model of Universe Expansion

30.1 Critical Density

In its observational form, Hubble's law relates the redshift z of galaxies to their distance d,

$$z = H_0 d/c, \tag{30.1}$$

where c is the speed of light, and the Hubble constant H_0 has units of inverse time. For nearby galaxies, the Doppler formula implies that the redshift is just linearly proportional to the speed of recession v,

$$z = \frac{\Delta\lambda}{\lambda} = \frac{v}{c}, \tag{30.2}$$

which when applied to Eq. (30.1) gives the velocity form of Hubble's law,

$$v = H_0 d. \tag{30.3}$$

This form has the simple and obvious interpretation that we currently live in an expanding universe. Indeed, if H_0 is strictly taken to be constant, then its inverse defines the "Hubble time" (see Eq. (27.4)),

$$t_H \equiv \frac{1}{H_0} \approx \frac{10\,\text{Gyr}}{h_0} \; ; \; h_0 \equiv \frac{H_0}{100(\text{km/s})/\text{Mpc}}, \tag{30.4}$$

which effectively marks the time in the past since the expansion began. As such, this Hubble time provides a simple estimate of the age of the universe since the Big Bang, with the latter equality giving the age in Gyr in terms of the scaled Hubble parameter $h_0 \equiv H_0/(100\,(\text{km/s})/\text{Mpc})$.

But, more realistically, one would expect the universe expansion to be slowed by the persistent inward pull of gravity from its matter, much the way that an object launched upward from Earth is slowed by its gravity. Indeed, a key question is whether gravity might be strong enough to stop and even reverse the expansion, much as occurs when an object is launched with less than Earth's escape speed.

For two points separated by a distance $d = r$, the relative speed is set by the Hubble law $v = H_0 r$. The associated kinetic energy per unit mass associated with the universe's expansion is thus

$$KE = \frac{v^2}{2} = \frac{H_0^2 r^2}{2}. \tag{30.5}$$

For a uniform density ρ, the total mass in a sphere of radius r centered on the other point is just $M(r) = 4\pi r^3 \rho/3$. The associated gravitational potential energy per unit mass is thus

$$PE = \frac{GM(r)}{r} = \frac{4\pi}{3}G\rho r^2.$$

(30.6)

Setting $KE = PE$, we can readily solve for the present-day critical density needed to just barely halt the expansion,

$$\rho_{co} = \frac{3H_o^2}{8\pi G} = 1.87 \times 10^{-29} h_o^2 \approx 9.2 \times 10^{-30} \frac{g}{cm^3}; \quad H_o \approx 70(km/s)/Mpc.$$

(30.7)

The last evaluation applies for the current observationally inferred, best value of the Hubble constant, $H_o \approx 70\,(km/s)/Mpc$, i.e., $h_o = 0.7$. Note that the arbitrary distance r has canceled out, demonstrating that this critical-density condition (Eq. (30.7)) applies to the expansion as a whole. If the universe has a present-day density $\rho_o > \rho_{co}$, the expansion will be stopped and even reversed, as we will now quantify by solving for the level of this gravitational deceleration.

30.2   Gravitational Deceleration of Increasing Scale Factor

Building upon this notion of gravitationally induced slowing of a critically expanding universe, let us now consider the net deceleration for a universe with a *noncritical* density ρ that is still uniform in space, but changes in time due to the expansion. Writing the present-day distance as $d = r(t = 0) \equiv r_o$ and the present-day density as $\rho_o \equiv \rho(t = 0)$, then because volume changes with expansion radius as r^3, we can see from mass conservation that the density at other times must scale as $\rho(t) = \rho_o r_o^3/r(t)^3$. The self-gravity of this mass density then causes a *deceleration* of the expansion,

$$\ddot{r}(t) = -\frac{GM(r)}{r^2} = -\frac{4\pi}{3}G\rho r = -\frac{4\pi G\rho_o r_o^3}{3r^2},$$

(30.8)

where the dots represent time differentiation, i.e., $\dot{r} \equiv dr/dt$ and $\ddot{r} \equiv d^2r/dt^2$.

For convenience, let us next introduce a changing spatial *scale factor* for this universal expansion,

$$R(t) \equiv \frac{r(t)}{r_o},$$

(30.9)

so that, by definition, $R_o \equiv R(t = 0) = 1$. Hubble's law then gives, for the present-day expansion rate,

$$\dot{R}_o \equiv \dot{R}(t = 0) = \frac{\dot{r}(t = 0)}{r_o} = \frac{v}{d} = H_o.$$

(30.10)

The deceleration equation (30.8) can thereby be written in the scaled form,

$$\ddot{R}(t) = -\frac{4\pi G \rho_0}{3R^2} = -\frac{\rho_0}{\rho_{co}}\frac{H_o^2}{2R^2} = -\Omega_m \frac{H_o^2}{2R^2}, \qquad (30.11)$$

where the very last equality defines the critical-density mass[1] fraction in the present universe,

$$\Omega_m \equiv \frac{\rho_0}{\rho_{co}}. \qquad (30.12)$$

Multiplying both sides by the expansion rate $\dot{R}(t)$, we can obtain a first integral of Eq. (30.11),

$$\dot{R}^2 = \frac{\Omega_m H_o^2}{R} - k = \frac{\Omega_m H_o^2}{R} + (1 - \Omega_m)H_o^2, \qquad (30.13)$$

where k is an integration constant, evaluated in the latter equality by using Eq. (30.10).

Noting that the Hubble constant here provides the scale for the time derivative, we can simplify the notation by measuring time in units of the Hubble time, and so making the substitution $t/t_H = H_o t \to t$. In such "Hubble units," Eq. (30.13) takes the simpler form with the Hubble constant replaced by unity ($H_o \equiv 1$),

$$\boxed{\dot{R}^2 = \frac{\Omega_m}{R} + (1 - \Omega_m).} \qquad (30.14)$$

Note then that, in addition to our original definition $R_o \equiv R(t = 0) = 1$, we now also have, in these units, $H_o \equiv \dot{R}_o \equiv \dot{R}(t = 0) = 1$.

The behavior of the expansion solution $R(t)$ depends on the critical density fraction Ω_m, as delineated in the following subsections, and plotted[2] in Figure 30.1. These solutions $R(t)$ versus t are computed by *inverting* the integral function for the time,

$$t(R) = \int_1^R \frac{dr}{\sqrt{\Omega_m/r + 1 - \Omega_m}}, \qquad (30.15)$$

where the lower bound of the integral at $R(t = 0) = 1$ was chosen so that this time is measured from the present $t = 0$, with any smaller $R < 1$ thus occurring in the past, $t < 0$. Equation (30.15) can be integrated analytically, but except for some special cases noted below, the full mathematical forms are quite complicated, and thus not obviously very instructive.

30.2.1 Empty Universe, $\Omega_m = 0$

The simplest case is that of an "empty" universe, $\Omega_m = 0$, representing the limit in which the mass density is too small to induce much gravitational deceleration. We then find that the expansion rate is constant, with $\dot{R} = 1$, which can be readily integrated,

---

[1] Note that this includes the total mass contributing to gravitational attraction, including both ordinary baryonic matter and dark matter.

[2] The relevant solutions here are those with no "cosmological constant" term, $\Omega_\Lambda = 0$; see Chapter 31 for the meaning of models with $\Omega_\Lambda \neq 0$.

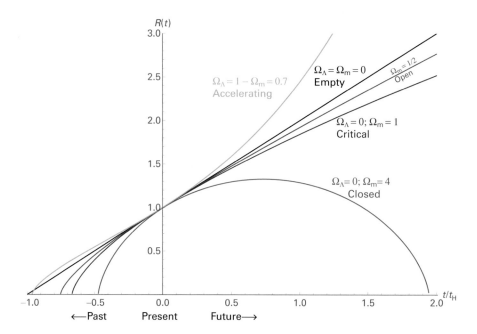

Figure 30.1 Cosmological scale factor R plotted versus time t in units of the Hubble time $t_{\rm H} \equiv 1/H_{\rm o}$, ranging from the past ($t < 0$), through the present ($t = 0$), to the future ($t > 0$), for various combinations for matter-critical density fraction $\Omega_{\rm m}$ and energy density fraction from the cosmological constant, Ω_Λ.

together with the boundary condition $R(t = 0) = 1$, to give a uniformly expanding scale factor that just increases linearly with time,

$$R(t) = 1 + t. \tag{30.16}$$

This case is illustrated by the straight black line in Figure 30.1.

30.2.2   Critical Universe, $\Omega_{\rm m} = 1$

Another case allowing simple integration is that of a critically dense universe, $\Omega_{\rm m} = 1$, for which Eq. (30.14) gives the expansion rate,

$$\dot{R} = R^{-1/2}. \tag{30.17}$$

Upon integration with the boundary condition $R(t = 0) = 1$, this gives the solution

$$R(t) = \left(1 + \frac{3}{2}t\right)^{2/3}. \tag{30.18}$$

This solution thus still expands forever, but approaches a vanishing rate, $\dot{R} \to 0$ as $t \to \infty$. It is illustrated by the purple curve in Figure 30.1.

30.2.3 Closed Universe, $\Omega_m > 1$

For a still-higher density fraction, $\Omega_m > 1$, the self-gravity can *halt* and *reverse* the expansion. From Eq. (30.14) the zero expansion rate $\dot{R} = 0$ occurs at a maximum scale factor,

$$R_{max} = \frac{\Omega_m}{\Omega_m - 1}. \tag{30.19}$$

As the universe thus eventually closes back on itself, this is known as a "closed" universe. It is illustrated by the red curve in Figure 30.1.

30.2.4 Open Universe, $\Omega_m < 1$

Finally, for subcritical density, the expansion again continues forever, but now with a nonzero asymptotic rate, given by taking $R \to \infty$ in Eq. (30.14),

$$\dot{R}_\infty = \sqrt{1 - \Omega_m}, \tag{30.20}$$

which implies that today's Hubble constant H_o would shrink by this factor $\sqrt{1 - \Omega_m}$ in the distant future. This is known as an "open" universe, illustrated by the blue curve in Figure 30.1.

30.3 Redshift versus Distance: Hubble Law for Various Expansion Models

Let us next consider how these various theoretical models for the universe connect with the observable redshift that indicates its expansion. Up to now, we have considered this redshift to be the result of the Doppler effect associated with distant galaxies receding from us at a speed that increases with distance, giving the speed–distance form of the Hubble law (Eq. (30.3)).

But an alternative, indeed more general and physically more appropriate perspective, is that this redshift is actually just a consequence of the *expansion of space itself!*

Recall that the basic definition of redshift is given in terms of the difference $\Delta\lambda = \lambda_{obs} - \lambda_{em}$ between the observed wavelength λ_{obs} and the originally emitted wavelength λ_{em},

$$z(d) \equiv \frac{\Delta\lambda}{\lambda_{em}} = \frac{\lambda_{obs}}{\lambda_{em}} - 1 = \frac{1}{R(t = -d/c)} - 1. \tag{30.21}$$

The last equality here follows directly from the definition of the scale factor R as the ratio of a length (here the emitted wavelength) at some remote time (set here by the light travel time $t = -d/c$ to the emitting object at distance d) to that observed at the present time.

For the simple case of a constantly expanding, empty universe, $R = 1 + H_0 t = 1 - H_0 d/c$, giving then for the distance of an object with given redshift,

$$d(z) = d_H \left(1 - \frac{1}{1+z} \right), \qquad (30.22)$$

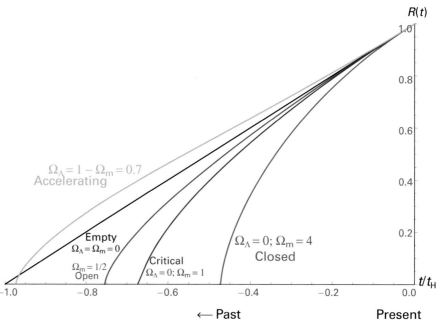

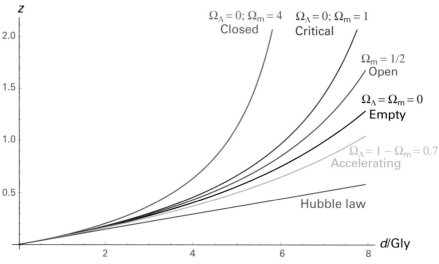

Figure 30.2 Top: Same as Figure 30.1, but focusing only on past times, $t < 0$. Bottom: Associated observable redshift z versus distance d, measured in Giga light years (Gly), assuming the current best estimate for Hubble constant $H_0 \approx 70$ (km/s)/Mpc, giving a Hubble time $t_H = 1/H_0 = 14.6$ Gyr.

where $d_H \equiv c/H_o$ defines the *Hubble distance*, i.e., the distance that light travels in a Hubble time $t_H \equiv 1/H_o$.

For past times that are small compared with the Hubble time, $-t \ll 1/H_o$, Taylor expansion gives $R(t) \approx R(t = 0) + \dot{R}_o t = 1 - H_o d/c$. When applied to Eq. (30.21), with further first-order expansion of the inverse binomial, this gives a simple linear Hubble law for distances small compared with the Hubble distance $d_H \equiv c/H_o$,

$$z(d) \approx \frac{1}{1 - H_o d/c} - 1 \approx \frac{H_o d}{c} = \frac{d}{d_H}; \quad d \ll d_H \equiv c/H_o, \qquad (30.23)$$

thus recovering the standard linear Hubble law (Eq. (30.1)).

But because the redshift depends on the *inverse* of the scale factor, for distances that are not small compared to the Hubble distance, the redshift-versus-distance relation becomes distinctly nonlinear, even for the linear expansion $R = 1 - d/d_H$ solution that applies for an empty universe with $\Omega_m = 0$. (See the black curve in the lower panel of Figure 30.2.)

For the same selection of expansion models as in Figure 30.1, Figure 30.2 compares plots of the scale factor R for past times $t < 0$ (top) with the associated variation of redshift z versus distance d (bottom). Note that, as implied by the expansion (30.23), all the models converge to the simple linear Hubble law (purple line) at modest distances, $d \ll d_H = c/H_o$. But for the inferred Hubble constant $H_o \approx 70\,\text{(km/s)/Mpc}$, giving a Hubble time $t_H \approx 14.6\,\text{Gyr}$ – which sets the *slope* of that initial line – we see that, at distances beyond $\sim 2\,\text{Gly}$, these models each start to deviate significantly from this linear Hubble law.

This deviation is greatest for the closed universe, but because of the inverse relation between redshift z and scale factor R, even the empty universe ($\Omega_m = 0$), with constant rate of expansion ($\dot{R}(t) = H_o$), shows a substantial deviation from the linear Hubble law for distances beyond about 2 Gly.

30.4   Questions and Exercises

Quick Questions

1. What is the age (in Gyr) of an "empty" universe with constant expansion and Hubble constant $H_o = 67$ (km/s)/Mpc?
2. What is the age (in Gyr) of a "critical" universe ($\Omega_m = 1$) and Hubble constant $H_o = 67$ (km/s)/Mpc?

Exercises

1. *Critical universe redshift*
   Consider a critical universe $\Omega_m = 1$ without dark energy ($\Omega_\Lambda = 0$) and a local Hubble constant equal to the currently inferred best-value $H_o \approx 70\,\text{(km/s)/Mpc}$.
   a. Derive a formula for redshift z versus distance d (in Mpc).
   b. Show that, for small distances $d \ll c/H_o$, this recovers the simple linear Hubble law $cz = H_o d$.

    c. Compute the time since the Big Bang (in Gyr).

    d. Compare this time to the age of a globular cluster with a main-sequence turnoff at luminosity $L_{to} = 0.75 L_{\odot}$.

    e. What does this say about the viability of this as a model for our universe? What about closed-universe models with $\Omega_m > 1$? (Assume the above Hubble-constant measurement is accurate, and that there is no dark energy.)

2. *Empty universe*

    Next consider the case of an effectively "empty" universe with $\Omega_m = \Omega_{\Lambda} = 0$, that is again expanding with a locally measured Hubble constant $H_o \approx 70\,(\text{km/s})/$ Mpc. Repeat parts a–d of the previous exercise for an empty universe. What does the result in part d here say absout the formal viability of this as a model for our universe?

3. *Empty versus critical universe*:

    a. For the above empty universe model, invert the formula for $z(d)$ to derive an expression for distance as a function of redshift z. For this use the notation $d_0(z)$, where the subscript "0" denotes the null value of Ω_m.

    b. If a distance measurement is accurate to 10 percent, at what minimum redshift z_0 can one observationally distinguish the redshift versus distance of an empty universe from a strictly *linear* Hubble law $d = cz/H_o$.

    c. Using the above results from Exercise 1a, now derive an analogous distance versus redshift formula $d_1(z)$ for the critical universe with $\Omega_m = 1$ (and $\Omega_{\Lambda} = 0$).

    d. Again, if a distance measurement is accurate to 10 percent, at what minimum redshift z_1 can one observationally distinguish the redshift versus distance of such a critical universe from a strictly linear Hubble law.

    e. Finally, again with a distance measurement accurate to 10 percent, at what minimum redshift z_{10} can one observationally distinguish the redshift versus distance of a critical universe from an empty universe?

31 Accelerating Universe with a Cosmological Constant

31.1 White-Dwarf Supernovae as Distant Standard Candles

To test which of these models applies to our universe, one needs to extend redshift measurements to large distances, out to several Gly. As long as an object is bright enough to show detectable spectral lines, measurement of redshift is straightforward, with, for example, quasars showing redshifts up to $z \approx 6.5$.

But it is much more difficult to get an *independent* measurement of distance for suitably remote objects. The most successful approach has been to use white-dwarf supernovae (WD-SN, also known as type Ia, or SN Ia) as very luminous *standard candles*. Because these supernovae all begin with similar initial conditions, triggered when accretion of matter from a companion pushes a white-dwarf star beyond the Chandrasekhar mass limit $M \approx 1.4 M_{\odot}$, they tend to have a quite similar peak luminosity, $L \approx 10^{10} L_{\odot}$, corresponding to an absolute magnitude $M \approx -20$. From the observed peak flux F or apparent magnitude m, one can then independently infer the distance $d = \sqrt{L/4\pi F} = 10^{1+(m-M)/5}$ pc $\approx 10^{5+m/5}$ pc. Thus, for example, observational surveys with a limiting magnitude $m \approx +20$ can detect WD-SN out to distance of $d \lesssim 10^9$ pc = 1 Gpc.

When combined with spectral measurements of the associated redshift z, the data from such white-dwarf supernovae place data points in a diagram of z versus d, such as that shown in Figure 30.2. For modest distances, $d \lesssim 1 - 2$ Gly < 1 Gpc, the slope of a best-fit line thus provides a direct measurement of the Hubble constant, H_0. But to measure deviations from a linear Hubble law, and so determine which of the above deceleration models best matches the actual universe, there was a concerted effort during the 1990s to discover and observe such supernovae in galaxies at greater and greater distances and redshifts. And as points were added at larger distances, they did indeed show the expected trend above this linear Hubble law, marked by the purple line in Figure 30.2.

But in one the greatest surprises of modern astronomy, and indeed of modern science, such data points were found to generally lie *below* the black curve that represents a nearly-empty universe, with a *constant* expansion rate $\dot{R} = H_0$. This immediately *rules out all the decelerating* models that lie above this black curve representing constant-rate expansion.

Instead it implies that the expansion of the universe must be *accelerating!*

31.2 Cosmological Constant and Dark Energy

For the universe's expansion to be accelerating, it is required that, in opposition to the attractive force of gravity, there must be a positive, repulsive force that pushes galaxies apart. Ironically, in an early ($\sim$1917) application of his General Relativity theory, Einstein had posited just such a universal repulsion term – dubbed the "cosmological constant," and traditionally denoted Λ. This was introduced to balance the attractive force of gravity, and so allow for a static, and thus eternal, model of the universe, which was the preferred paradigm at that time. Then, after Hubble's discovery that the universe is not static but expanding, Einstein completely disavowed this cosmological-constant term, famously calling it "his greatest blunder."

But nowadays, with the modern discovery that this expansion is actually *accelerating*, the notion of something akin to the cosmological constant has been resurrected. The full physical bases and origin are still quite unclear, but the effect is often characterized as a kind pressure or tension of spacetime itself, with associated mass–energy density, dubbed *dark energy*, parameterized in terms of the fraction Ω_Λ of the critical mass–energy density $\rho_{co}c^2$.

While a rigorous discussion requires a general relativistic treatment beyond the scope of this book, within the above simplified Newtonian model for time evolution of the universe's scale factor R, this dark energy can be heuristically accounted for by adding a *positive* term to the right-side of Eq. (30.11),

$$\ddot{R}(t) = -\frac{4\pi G \rho_o}{3R^2} + \frac{\Lambda R}{3} = -\Omega_m \frac{H_o^2}{2R^2} + \Omega_\Lambda H_o^2 R. \tag{31.1}$$

Note that now the acceleration transitions from strongly *negative* in the early universe with small scale factor $R \ll 1$, to strongly *positive* in an older universe with large scale factor $R \gg 1$. The transition, with momentarily zero acceleration ($\ddot{R} = 0$), occurs at a scale factor

$$R_z = \left(\frac{\Omega_m}{2\Omega_\Lambda} \right)^{1/3}. \tag{31.2}$$

Again using the $\dot{R}$ integrating factor and setting $H_o \equiv 1$ to define time in terms of the Hubble time, we obtain a generalized first integral solution (cf. Eq. (30.14)),

$$\dot{R}^2 = \frac{\Omega_m}{R} + \Omega_\Lambda R^2 + (1 - \Omega_m - \Omega_\Lambda), \tag{31.3}$$

where we have again evaluated the integration constant by using the boundary conditions $R_o = \dot{R}_o = 1$.

In general relativity gravity is described in terms of the warping, or *curvature*, of spacetime.[1] In its application to cosmology, the value of the term in parentheses in Eq. (31.3) sets the overall curvature of the whole universe, with positive, negative, and zero values corresponding to curvatures that are similarly positive (like a sphere), negative (like a saddle), and zero (like a flat sheet). Figure 31.1 illustrates these cases.

---

[1] In relativity theory, space and time are combined into a coupled *spacetime*.

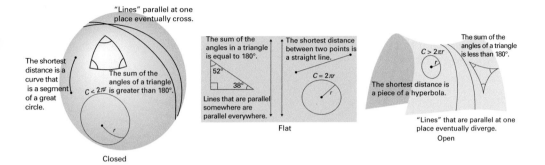

Figure 31.1 Illustration of three cases for curvature in ordinary 3D space, ranging from the positive curvature of a closed sphere, to the zero curvature of a flat surface, to the negative curvature with an open saddle. The annotations show how the different geometries lead to different properties for angles and distances.

31.3   Critical, Flat Universe with Dark Energy

As discussed below, there are strong theoretical arguments (e.g., from the theory of inflation; see Section 33.5) that the universe must be geometrically nearly *flat*, meaning then that the term in parentheses in Eq. (31.3) is very nearly zero. This, in turn, implies that the total energy density is very near the *critical* value, with $\Omega_{\rm m} + \Omega_\Lambda = 1$. Using this to eliminate Ω_Λ, we can again cast the range of possible models in terms of the single parameter $\Omega_{\rm m}$, with Eq. (31.3) reducing to

$$\dot{R}^2 = \frac{\Omega_{\rm m}}{R} + (1 - \Omega_{\rm m})R^2. \tag{31.4}$$

For a such a critical, flat universe with the various labeled values of $\Omega_{\rm m} = 1 - \Omega_\Lambda$, the top panel of Figure 31.2 plots the expansion rate $\dot{R}$ (in units of Hubble constant $H_{\rm o}$) versus scale factor R. Note that, for $\Omega_{\rm m} > 0$, the expansion starts very rapidly, but then declines to a minimum of order $H_{\rm o}$, and finally increases again for large R. The bold black curve is for the case $\Omega_{\rm m} = 0.3$, which as discussed below, is roughly the best-fit value for our universe. Note that, for much of the evolution of $0.2 \lesssim R < 1$, the model has $\dot{R} \approx H_{\rm o}$. The bottom panel shows that the associated age for such a case with $\Omega_{\rm m} = 0.3$ is very close to the Hubble time $t_{\rm H} = 1/H_{\rm o}$ that applies for the simple case of a constantly expanding universe $R(t) = 1 + H_{\rm o}t$.

   A simple example for the full solution is again for the case of a matter-empty universe, $\Omega_{\rm m} = 0$, for which we find $\dot{R} = \pm R$. Choosing the plus root to represent the observed case of expansion, we find

$$R(t) = e^{H_{\rm o}t} = e^{t/t_{\rm H}}. \tag{31.5}$$

Thus, in contrast to the previous case of constant expansion for an empty universe with $\Omega_{\rm m} = \Omega_\Lambda = 0$, for a dark-energy-dominated, flat universe with $\Omega_\Lambda = 1$, the expansion actually *accelerates exponentially*, with an e-fold increase each Hubble time!

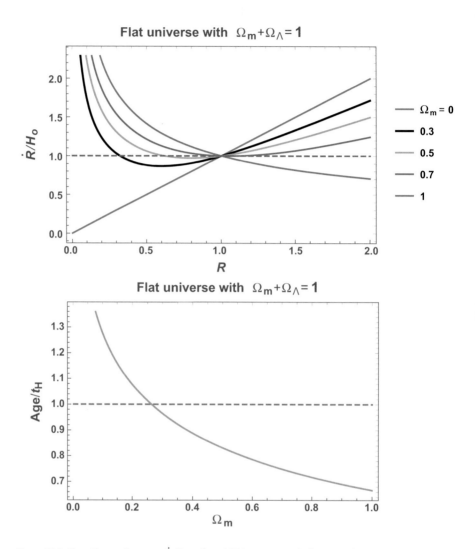

Figure 31.2 Top: Expansion rate $\dot{R}$ (in units of H_o) versus scale factor R for a flat universe with the various labeled values of $\Omega_m = 1 - \Omega_\Lambda$. Bottom: The age of a flat universe (in Hubble time $t_H \equiv 1/H_o$) plotted versus $\Omega_m = 1 - \Omega_\Lambda$.

In fact, even for the more general case with $0 < \Omega_m < 1$, note that, as R increases, the rate again becomes dominated by the second (cosmological acceleration) term in Eq. (31.4), implying $\dot{R} \sim +R$ and thus again an exponential expansion at large times!

The full solution can again be obtained by inverting the time integral solution, now giving for the time t_{BB} since the Big Bang (cf. Eq. (30.15)),

$$t_{BB}(R) = \int_0^R \frac{dr}{\sqrt{\Omega_m/r + (1 - \Omega_m)r^2}}, \tag{31.6}$$

which has the analytic form (see Question 1 at the end of this chapter),

$$t_{\text{BB}}(R) = \frac{2 \operatorname{arcsinh}\left[\sqrt{\frac{1-\Omega_{\text{m}}}{\Omega_{\text{m}}}} R^{3/2}\right]}{3\sqrt{1-\Omega_{\text{m}}}}, \tag{31.7}$$

where arcsinh is the inverse of the hyperbolic sine function, $\sinh(x) \equiv (e^x - e^{-x})/2$.

Applying this for $R = 1$, we obtain a simple analytic form for the universe's current age $t_{\text{age}} \equiv t_{\text{BB}}(R = 1)$ as a function of Ω_{m}. The bottom panel of Figure 31.2 plots t_{age} (in t_{H}) versus Ω_{m}. Note that for $\Omega_{\text{m}} = 0.3$, the associated age is $t_{\text{age}} = 0.964$, so very close to the one Hubble time that applies for the simple case of a constantly expanding universe $R(t) = 1 + t$.

Leaving details to Question 2 at the end of this chapter, this solution (Eq. (31.7)) can be readily inverted. When cast in terms of the time $t \equiv t_{\text{BB}} - t_{\text{age}}$ relative to the present, and normalized to give $R(t = 0) = 1$, the result is

$$R(t) = \left(\frac{\Omega_{\text{m}}}{1-\Omega_{\text{m}}}\right)^{1/3} \sinh[(3/2)\sqrt{1-\Omega_{\text{m}}}\,(t_{\text{age}}+t)]^{2/3}. \tag{31.8}$$

Equation (31.8) provides an analytic form for the time variation of the scale factor for the currently favored model of a flat universe with dark energy, for *any* presumed value of the matter density, $\Omega_{\text{m}} = 1 - \Omega_\Lambda$. Figure 31.3 plots R versus $t_{\text{BB}} = t + t_{\text{age}}$ for various Ω_{m}.

In terms of the present-day-based time t and associated redshift $z = -1 + 1/R$ versus distance $d = ct$, the green curves in Figures 30.1 and 30.2 plot this solution

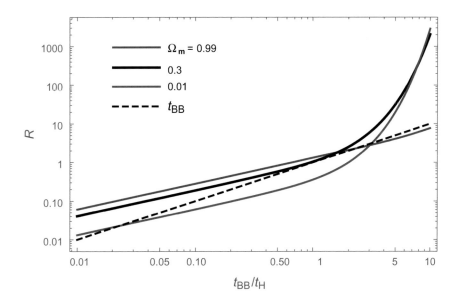

Figure 31.3 Scale factor R versus time t_{BB} since the Big Bang, for flat universe with $\Omega_{\text{m}} = 1 - \Omega_\Lambda$, plotted on a log–log scale for the most-favored value $\Omega_{\text{m}} = 0.3$ (black curve), as well as $\Omega_{\text{m}} = 0.01$ (red) and 0.99 (blue).

for the case $\Omega_m = 0.3$ and thus $\Omega_\Lambda = 0.7$, which turns out to best fit the SN data (as well as other constraints from fluctuations in the Cosmic Microwave Background, CMB). This implies that the combination of ordinary and dark matter makes up only about 30 percent of the mass–energy density of the universe, with the other about 70 percent in the form of this mysterious dark energy!

Inspection of Figures 30.1 and 30.2 shows that the green curves for this dark-energy model are actually not too different from the basic black curves, which represent the very simple "empty universe" model without *any* kind of matter-energy. In the absence of any forces, this model gives a simple "coasting" solution, with scale factor $R(t) = 1 + H_0 t$. Its rough agreement with the dark-energy model means that dark energy can be roughly thought of as providing an outward pressure that approximately cancels the inward attraction from gravity. Since the net force is nearly zero, the dark-energy solution is also nearly coasting, $R(t) \approx 1 + H_0 t$, at least for the universe *up to its present age*.

But recall that the first term on the right side of Eq. (31.4), which represents the inward pull of gravity, declines as $1/R$ as the scale factor R gets large, whereas the second term, representing the cosmological constant, actually *increases quadratically* with increasing R. Thus quite unlike a truly empty, coasting universe, for which the scale factor just *increases linearly* in time, $R(t) = 1 + H_0 t$, a universe with a nonzero cosmological constant $\Lambda > 0$ will eventually grow *exponentially*.

31.4   The "Flatness" Problem

One general puzzle for any model of the universe is that having the universe be nearly flat today, with total $\Omega_o = \Omega_m + \Omega_\Lambda \approx 1$, requires that it must have been even *much* flatter, with $\Omega(t)$ much closer to unity, in the past ($t < 0$).

To see this, let us write write the constant total energy per unit mass E_{tot} of the expanding universe in terms of the sum of its associated kinetic and potential energy components,

$$v^2 - \frac{2GM(r)}{r} = H^2 r^2 - \frac{8\pi G \rho r^2}{3} = 2E_{tot}. \tag{31.9}$$

By dividing by the second term in the middle expression, this can be recast into the form

$$\frac{1 - \Omega(t)}{\Omega(t)} = \frac{\rho_{co}}{R(t)^2 \rho(t)} \frac{1 - \Omega_o}{\Omega_o}, \tag{31.10}$$

where $\Omega(t) \equiv \rho(t)/\rho_c(t)$ is the critical-density fraction at some earlier time t, with $\rho_c(t) \equiv 3H(t)^2/8\pi G$ a generalization of Eq. (30.7) to define the critical density at this time when the Hubble expansion rate now has a general time variation $H(t)$. On the right-hand side, the total energy and other constants have thus been cast in terms the critical density Ω_o in the current-day universe. If this Ω_o differs from unity by some small fraction, say $|1 - \Omega_o| \approx 0.01$ (i.e., 1 percent) in the current-day universe then, in the earlier universe, the difference is smaller by a factor

$$|1 - \Omega(t)| \approx \frac{0.01}{R(t)^2 \rho(t)/\rho_{co}} \approx 0.01\, R(t); \quad 10^{-4} < R < 1 \tag{31.11}$$

$$\approx 100\, R^2; \quad R < 10^{-4} \tag{31.12}$$

The upper equality assumes a matter-dominated universe with density $\rho \sim 1/R^3$. But, as discussed in Sections 32.1 and 33.1, the *temperature* of the universe scales as $T \sim 1/R$; thus, in the early universe with $R < 10^{-4}$, the energy density was dominated by *radiation*, since radiation's energy density scales as $U_{rad} \sim T^4 \sim 1/R^4$, i.e., one higher factor of $1/R$ than the $\rho \sim 1/R^3$ scaling of matter. Extending back to very early times, we thus require $|1 - \Omega| \sim R^2 \to 0$, meaning then that any "initial" deviations from flatness had to have been extremely tiny.

If, instead, the initial Ω had been even slightly above unity, the fledgling universe would have recollapsed as a tiny, closed universe. Alternatively, if Ω had been even slightly below unity, the universe would have expanded at such a high rate that galaxies would not have had time to form. Overall, this fine tuning required to make $|1 - \Omega|$ initially very small is known as the "flatness" problem for reaching the kind of moderately expanding, mature universe we live in today. Section 33.5 discusses how the concept of *cosmic inflation* helps resolve this flatness problem.

But let us next consider further the temperature history and associated properties of the universe extending back to such early times of a "Hot Big Bang."

31.5   Questions and Exercises

Quick Questions

1. Confirm the analytic integral result in Eq. (31.7) by taking the derivative dt_{BB}/dR and comparing this with the integrand of Eq. (31.6).

2. Confirm that inversion of the integral solution (Eq. (31.6)) results in the solution (Eq. (31.8)) for $R(t)$.

3. Show that, for $x \ll 1$, $\sinh(x) \approx x$.

4. For ages $t_{BB} \ll t_H$ in the ΛCDM model, show that:

$$R(t_{BB}) \approx (9\Omega_m/4)^{1/3}(t_{BB}/t_H)^{2/3}.$$

5. Using Eq. (31.7), determine (e.g., by trial and error) the value of Ω_m for which $t_{age} = t_H$ in a model of a flat universe with a cosmological constant.

6. Referring to Figure 31.3, discuss how the red and blue curves relate to the simple special solutions derived for a flat, empty universe with dark energy, and a critical universe with no dark energy.

Exercises

1. *Fine tuning of gravitational escape*
   Suppose one launches an object with a speed $V = V_{esc}(1 - \delta)$ that is slightly less than the escape speed $V_{esc} = \sqrt{2GM/R}$ from a surface radius R.
   a. How finely tuned must δ be for the object to reach maximum radii $r/R = 10, 100, 10^{10}, 10^{100}$?
   b. Discuss how this is related to the fine tuning needed for the universe to reach its current scale factor $R = 1$ from initial values $R \ll 1$.

2. *Acceleration rate of universe*
   For an empty universe with dark energy, Eq. (31.5) gives the change in the scale factor R with time t.
   a. Derive an expression for the associated *acceleration* in the scale factor, $\ddot{R}$.
   b. Converting to a dimensional scale $r = Rd_H$, where $d_H = c/H_o$ is the Hubble length, next write this in terms of dimensional acceleration, $\ddot{r}$.
   c. For a Hubble constant $H_o = 70\,(km/s)/Mpc$, evaluate $\ddot{r}$ in MKS and CGS units at one Hubble time t_H after the Big Bang.
   d. How does this compare with the Earth's acceleration of gravity g_e?
   e. At what time in the future will $\ddot{r} = g_e$? Give this in both Gyr and Hubble times.

3. *Flat universe with dark energy*
   Consider the analytic expression (Eq. (31.8)) for the time variation of the scale factor $R(t)$ in a flat universe with dark energy.
   a. Confirm explicitly that $R(t = 0) = 1$ by plugging Eq. (31.7) into Eq. (31.8) for the present-day time $t = 0$.
   b. Show that, for small times $t_{BB} \gtrsim 0$ after the Big Bang, $R \sim t_{BB}^{2/3}$.
   c. Show that, for large times in the future $t_{BB} \gg 1$, the growth in R become exponential, and give an explicit expression for its e-folding time t_e in terms of Ω_m and t_H.

4. *Inverted-pendulum analog for dark energy*
   Consider an inverted pendulum in which a mass m is balanced vertically on a rigid but massless rod of length ℓ in the presence of a downward gravity $-g$.
   a. For small angle displacements $\theta \ll 1$, show that $\ddot{\theta} = +(g/\ell)\theta$.
   b. How does this differ from the usual case of a downwardly suspended pendulum?
   c. For the inverted case, obtain solutions for $\theta(t)$ given initial conditions $\theta(t = 0) = 0$ and $\dot{\theta}(t = 0) = \theta_o$.
   d. Similarly for the matterless dark-energy model of Eq. (31.1) with $\Omega_m = 0$, obtain solutions for $R(t)$ given $R(t = 0) = 0$ and $\dot{R}(t = 0) = H_o$.
   e. What are the inverted-pendulum analogs to R, H_o, and Ω_Λ?

32 The Hot Big Bang

32.1 The Temperature History of the Universe

The smaller scale factor of the past universe clearly means its overall averaged density was higher than it is today, since $\rho \sim 1/R^3$. But what might we conclude about the overall *temperature* history of the universe? In the present-day universe the temperature of individual structures varies widely, e.g., from millions of kelvin in the interiors of stars, to just a few degrees above absolute zero in cold giant molecular clouds; so it might seem absurd to even speak of a single temperature for the whole universe.

But if we go back in time before all this structure, when the density of the universe was much higher and much smoother, there was a kind of *thermal equilibrium* that led to a quite well-defined characteristic temperature. Intuitively, we can expect that in the smaller, more compressed, and thus much denser, early universe, the temperature should also be correspondingly much higher.

This dense, hot early universe emitted radiation like a blackbody, and a key property of the associated Planck function is that the overall form of fractional energy distribution over wavelength depends only on the product λT. Thus the redshift of an observed versus emitted wavelength – by a factor $\lambda_{obs}/\lambda_{em} = 1 + z$ – can just be accounted for by reducing the observed versus emitted temperature – by a factor $T_{obs}/T_{em} = 1/(1 + z)$. And since $1 + z = 1/R$, this then implies that this radiation temperature of the universe just increases with the inverse of the scale factor, $T \sim 1/R$.

For example, at a redshift of $z \approx 1000$, corresponding to a scale factor $R \approx 10^{-3}$, the temperature of the universe was about as hot as the surface of a relatively cool star, $T \approx 3000\,\text{K} \approx T_{\odot}/2$. And much as a star, this hotter early universe emitted radiation according to the blackbody function $B_{\lambda}(T)$ for that temperature, with an original emitted spectrum that had its peak at a wavelength $\lambda_{max} = 500\,\text{nm}\,T_{\odot}/T \approx 1\,\mu\text{m}$.

But in the present-day universe this radiation should be *redshifted* by a factor $z = 1/R - 1 \approx 10^3$, with a corresponding peak wavelength in the *microwave* region (like in your microwave oven), $\lambda_{max} \approx 10^3\,\mu\text{m} \approx 1\,\text{mm}$. Moreover, in contrast to the directed "outward" emission from a star, this cosmic radiation was emitted *isotropically* (equal in all directions), and so would be observed today from all directions in the sky, as what is known as the Cosmic Microwave Background (CMB).

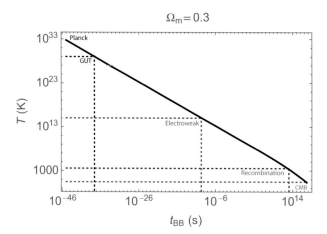

Figure 32.1 Temperature T (in K) versus time t_{BB} since the Big Bang, for a flat universe with $\Omega_m = 1 - \Omega_\Lambda = 0.3$, modified to account for the dominance of radiation at early times. The dotted lines mark the temperatures and times for various key epochs, as labeled. (The temperature plotted here uses the bridging form $R_b(t)$ given in part f of Exercise 4 at the end of Chapter 33.)

As discussed in the next section, this CMB fits closely to a blackbody with temperature $T_{cmb} \approx 2.7$ K. The upshot then is that the temperature history of the universe follows the simple form

$$T(t) = \frac{T_{cmb}}{R(t)} = T_{cmb} (1 + z). \tag{32.1}$$

Figure 32.1 plots the time evolution of temperature $T \sim 1/R$ for a model that accounts both for the role of dark energy in the later universe, and the dominance of radiation at early times (see Section 33.1 and Exercise 4 at the end of Chapter 33). The dotted lines represent temperatures and times ranging from the present day with $T_{cmb} = 2.7$ K (red) to the Planck time with $T_{Planck} \approx 10^{33}$ K, at which gravity melds with quantum physics, representing the limit of our present-day physics. This and the intervening epochs for recombination (purple), electroweak unification (blue), and Grand Unified Theory (GUT; black) unification are discussed in Chapter 33.

32.2   Discovery of the Cosmic Microwave Background

Early proponents of this "Hot-Big-Bang" model – most notably Robert Dicke of Princeton – actually predicted such a CMB before it was detected, rather serendipitously, in 1965, by two engineers named Penzias and Wilson from Bell Labs. They were actually just trying to reduce the persistent noise that was inherent in the radio receivers they were developing for communications, in some ways the predecessors of microwave antennae used for mobile phones today. After working hard to reduce

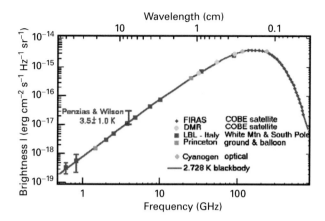

Figure 32.2 Sky brightness of CMB versus frequency (bottom axis) or wavelength (top axis) on a log-log scale, showing the nearly perfect fit of data from COBE and other measurements to a Planck blackbody function of temperature $T_{cmb} = 2.728$ K (magenta curve). Credit: NASA/COBE.

electronic and other[1] possible sources of static, they eventually concluded the noise was actually coming from the sky. Noting moreover that it was constant over both night and day, with a uniform brightness over the whole sky (and not, for example, concentrated along the equator, ecliptic, or the plane of the Milky Way), they, with some help from reading an unpublished preprint by Dicke and his colleagues, identified it as the predicted CMB. This momentous discovery, which provided striking confirmation of the Hot-Big-Bang model, eventually earned them (but not Dicke) a share of the 1978 Nobel Prize in Physics.

Subsequent observations have shown that the CMB is indeed isotropic to a very high precision, with fluctuations less than one part in 10 000 ($<10^{-4}$)! Moreover, as illustrated in Figure 32.2, it also follows both the form and absolute surface brightness[2] of the Planck blackbody function to a similarly high precision, with an inferred temperature $T_{cmb} = 2.728 \pm 0.001$ K. This can be considered as the present-day "temperature of our universe."

32.3 Fluctuation Maps from the COBE, WMAP, and Planck Satellites

Although the CMB appears isotropic and uniform down to levels $<10^{-4}$, the universe we live in today is very nonuniform, with large-scale structure, superclusters, galaxies, stars, and planets. Even with the extra mass from dark matter to enhance the mutual

---

[1] Including, they reported, from pesky avian deposits of "dielectric material" on the antennae.
[2] Recall that, in contrast to the flux from a localized source, surface brightness of an angularly resolved source does not decline with distance. Thus, once accounting for the redshift expansion of the universe in reducing the CMB temperature, the surface brightness of the CMB is the same today as what was emitted at the end of the recombinations era!

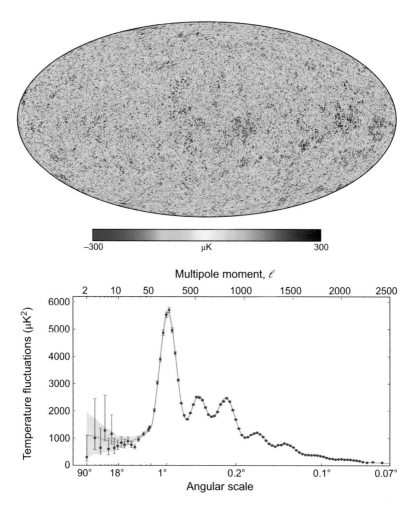

Figure 32.3 Top: Sky map of temperature fluctuations in the Cosmic Microwave Background (CMB), as measured by the Planck satellite. Bottom: Power spectrum of temperature fluctuations plotted versus angular scale (lower axis) or spherical harmonic multipole moment ℓ (upper axis). Credit: ESA and the Planck Collaboration.

gravitational attraction, any contraction to form this extensive structure still requires initial "seeds" in the form of small-amplitude fluctuations in local density. From simulation models for the formation of large-scale structure, it was predicted during the 1980s that the level of fluctuations needed would impart small fluctuations in the CMB at the level of a few parts per 100 000, i.e., a few times 10^{-5}, implying temperature fluctuations up to $\Delta T \lesssim 10^{-4} T_{cmb} \approx 300\,\mu K$.

Detecting such fluctuations thus became a major goal for observation and experiment. For ground-based observations it is very difficult to remove the effects of the Earth's atmosphere to a level that does not mask the predicted fluctuations, though there was some success, for example from a balloon-born experiment called *Boomerang* that circled the south pole. But the clearest results came from a series of orbiting satellites named COBE (COsmic Background Explorer, launched in 1989), WMAP (Wilkinson Microwave Anisotropy Probe, launched in 2001), and Planck (launched in 2009). COBE succeeded in measuring fluctuations at a level of about $200\,\mu\text{K}$, or $\lesssim 10^{-4}$, but its angular resolution was limited to large angular scales, >7 degrees. WMAP, and then Planck, greatly improved both the precision and the angular resolution, with Planck measuring fluctuations down to a precision of a few μK (i.e., $\Delta T / T \sim 10^{-6}$), at angular scales $<0,1$ degree.

The top panel of Figure 32.3 shows a full-sky map (in galactic coordinates, with galactic plane extending horizontally from the galactic center) of the CMB temperature fluctuations (in μK), as measured by the Planck satellite. The color range over $\pm 300\mu\text{K}$ represents relative fluctuations up to $\pm 300\,\mu\text{K} \sim \pm 10^{-4}\, T_{\text{cmb}}$, with red hotter and blue cooler.

The spatial power spectrum in the lower panel shows that these fluctuations occur over a range of angular scales, with main peak at about 1 degree. While the observed CMB comes from the last scattering during the recombination era, the fluctuations originate from processes before this era. Much as measurement of seismological waves generated in an earthquake provide information on the interior structure of the Earth, these measures of CMB fluctuation power peaks provide information on the prerecombination evolution of the universe, and place strong constraints for basic cosmological parameters.

In general, the analysis finds remarkably good overall agreement with predictions of the now-standard "ΛCDM" model of a universe, in which there is both "cold dark matter" (CDM) to spur structure formation, as wall as dark-energy acceleration that is well-represented by a cosmological constant Λ.

Specifically the Planck analysis quotes values $\Omega_b = 0.049$ for the fraction of ordinary (baryonic) matter, $\Omega_{\text{dm}} = 0.268$ for the fraction of dark matter, and $\Omega_\Lambda = 0.682$ for the fraction of dark energy; the sum, $\Omega_o \equiv \Omega_b + \Omega_{\text{dm}} + \Omega_\Lambda = 1$, is fully consistent with a flat universe.

It also inferred a value[3] $H_0 = 67.4 \pm 0.5$ (km/s)/Mpc for the Hubble constant, and 13.8 Gyr for the age of the universe.

32.4   Questions and Exercises

Quick Questions

1. What will the CMB temperature be when the universe is twice as big as it is now?
2. Compute the Hubble times t_H (in Gyr) for both the Planck and Cepheid values for the Hubble constant H_0. For a ΛCDM model with the Planck values for Ω_b and Ω_{dm}, what are the associated present-day ages, again in Gyr?

---

[3] This differs from the value $H_0 \approx 74$ (km/s)/Mpc inferred from astronomical observations based on Cepheid standard candles. The difference is several times the quoted error estimates of the two methods, and may suggest the operation of some new, not-yet-known physics.

Exercises

1. *Ice in the early universe?*
   a. At what redshift z was the CMB temperature 273 K?
   b. What is the associated scale factor R?
   c. For a simple constant expansion universe with $H_o = 70$ (km/s)/Mpc, what was the universe's age when it had this temperature?
   d. For a flat, dark-energy model with $\Omega_m = 0.3$, what was the universe's age when it had this temperature?
   e. If water existed at this time, could interstellar ice form?
   f. Finally, discuss whether water molecules could even exist at this time. Why or why not?

2. *Planck function*
   a. For a Planck function $B_\nu(T)$ at frequency ν for temperature T, show that the fractional distribution of energy in a given frequency interval ν and $\nu + d\nu$ – given by $B_\nu(T)d\nu/B(T)$ (where $B(T) = \sigma_{sb}T^4/\pi$ is the frequency-integrated emission given in Eq. (5.1)) – depends only on the dimensionless ratio, $h\nu/kT$.
   b. Similarly, for the wavelength form of the Planck function $B_\lambda(T)$, show that the fractional distribution of energy in wavelength λ depends only on the dimensionless ratio, $hc/\lambda kT$.

3. *Energy density of CMB versus starlight*
   a. For present-day CMB temperature $T \approx 3$ K, compute the CMB radiation energy density e_{cmb} (in erg/cm^3).
   b. What was the associated value at recombination, e_{rec}?
   c. For a simple model of the solar neighborhood of all being solar-type stars with a volume density of 1/pc^3, what is the associated local energy density e_* of their starlight?
   d. Now assuming this applies on average to galaxies such as our Milky Way which have a volume density of 1/Mpc^3, what is the associated local energy density e_{gal} of the galactic starlight? (Hint: First work out the volume filling fraction of galaxies.)
   e. Compare these values, and briefly comment on the physical relevance.

4. *Dipole temperature anisotropy of CMB*
   Observations of the CMB temperature shows a large-scale dipole variation, with opposite sides of the sky having a temperature difference $\Delta T = \pm 3.4 \times 10^{-3}$ K compared with the overall mean $T_{cmb} \approx 2.7$ K.
   a. If this is to be explained by the Doppler effect, what is the associated speed V_{dipole} relative to the overall universe.
   b. How does this compare to the orbital speeds V_e of the Earth around the Sun, and $V_\odot$ for the Sun around the Galaxy?
   c. Compute temperatures changes from V_e and $V_\odot$.
   d. Discuss briefly whether (or not) this violates Einstein's relativity principle that there is no special frame of reference.

33 Eras in the Evolution of the Universe

33.1 Matter-Dominated versus Radiation-Dominated Eras

Let us now consider the relative scalings of the radiation versus matter in the early universe (see Figure 33.1).

Because the energy density of radiation scales as $U(T) = a_{rad}T^4$ (where the radiation constant $a_{rad} \equiv 4\sigma_{sb}/c$; see Chapter 5), we can see that the radiative energy density has a scaling $U \sim T^4 \sim 1/R^4$ that is *steeper* (by one factor of $1/R$) than the density scaling, $\rho \sim 1/R^3$, of ordinary matter. For the present-day matter density $\rho_o = \Omega_m \rho_{co}$, the ratio of matter to radiation energy density is

$$\frac{\rho_o c^2}{U(T_{cmb})} = \frac{\Omega_m c^2 (3H_o^2/8\pi G)}{a_{rad}T_{cmb}^4} \approx 4.2 \times 10^4 \, h^2 \Omega_m \approx 6000, \tag{33.1}$$

where $h \equiv H_o/(100\,(\text{km/s})/\text{Mpc})$, and the last equality comes from applying the standard values $h \approx 0.7$ and $\Omega_m \approx 0.3$. Thus in our present-day universe *matter dominates over radiation* in terms of the associated mass–energy density.

However, since this ratio declines in direct proportion to the decreasing scale factor R, we find that, at a time with $R \approx 1/6000 \approx 10^{-4}$, when the redshift was approximately $z = 1/R - 1 \approx 10^4$, there is a transition to a higher density in radiation than matter, with earlier times with $R < 10^{-4}$ (and so $z > 10^4$) thus representing a *radiation-dominated* era.

Moreover, even though the mass–energy of the present-day universe is dominated by matter over radiation, it turns out that the *number* of CMB photons $n_\gamma(T_{cmb})$ actually greatly exceeds the number n_H of hydrogen atoms or protons. Since hydrogen is a mass fraction $X_H \approx 0.73$ of the ordinary matter that amounts to only about 5 percent of the critical density ρ_{co}, the present-day hydrogen number density is approximately

$$n_{H0} \approx \frac{0.73 \times 0.05 \rho_{co}}{m_p} = 1.8 \times 10^{-7} \, \text{cm}^{-3}. \tag{33.2}$$

The number of CMB photons can be estimated by dividing the energy density by an average photon energy, which for the CMB temperature is $\langle E \rangle \approx 3kT_{cmb} \approx 7 \times 10^{-4}$ eV. This gives

$$n_\gamma(T_{cmb}) \approx \frac{U(T_{cmb})}{\langle E \rangle} \approx \frac{a_{rad}T_{cmb}^4}{3kT_{cmb}} = 360 \, \text{cm}^{-3} \approx 2 \times 10^9 \, n_{H0}, \tag{33.3}$$

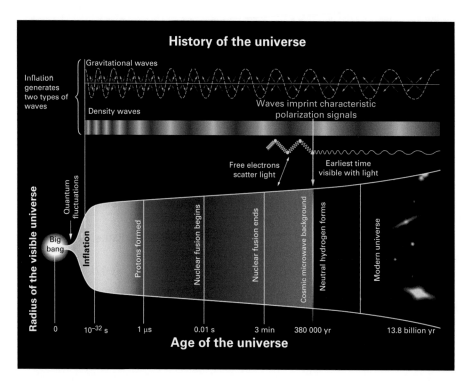

Figure 33.1 Illustration of different eras of the universe, back to the Big Bang. The upper part shows the generation of gravitational waves during the era of cosmic inflation, and how they could be detected through circular polarization imparted on radiation observed in the CMB. Credit: BICEP2 collaboration.

which shows that the photon *number* density is more than a billion times the proton density!

Moreover, since $n_\gamma \sim U(T)/T \sim T^3 \sim 1/R^3$, this photon density has the same $\rho \sim 1/R^3$ dependence on scale factor as the matter density. As such, the ratio $n_\gamma/n_p \sim 10^9$ thus remains roughly *constant* at this high value all the way back to the formation of the CMB, and indeed well into the radiation-dominated era.

As discussed in Section 33.3, this ratio plays an important role in the relative abundances of He and other light elements that form in the "era of nucleosynthesis," when $T \approx 10^9$ K.

But first, let us next consider more carefully the formation of the CMB, at the time when the temperature was $T \gtrsim 3000$ K, cool enough for electrons to recombine with protons, known thus as the "recombination era."

33.2   Recombination Era

In the early epochs of the Hot Big Bang, the temperature was so high that all the hydrogen was fully ionized, with proton number density $n_p = n_H$. Moreover, if for simplicity we neglect the contributions from helium, then overall charge neutrality requires an equal number of electrons, and so $n_e = n_p = n_H$. Because electrons can so readily scatter radiation, the photons of this era were effectively trapped, much as they are in the interior of a star. But as the universe cooled, the protons and electrons recombined to make neutral hydrogen, which is much less effective in absorbing or scattering radiation. The photons from this *recombination era* thus were suddenly free to propagate through the universe, becoming redshifted by its expansion to form the CMB we observe today.

We can model this CMB formation much as we model the emitted radiation from a star like our Sun. From the Eddington–Barbier relation of Section D.2 of Appendix D (see Eq. (D.3)) we see that the surface brightness at the center of the solar disk is set by the Planck function at about unit optical depth along that radial (i.e., $\mu = 1$) line of sight, $I_{obs} \approx B(\tau = 1)$. Analogously, the CMB surface brightness emitted at the recombination era can be derived from the electron-scattering optical depth. Integrating over the path of the photons, traveling at the speed of light c, from some past time ($t_p < 0$) to the present day ($t = 0$), this optical depth is given by

$$\tau_e(t) = \int_{t_p}^{0} \sigma_{Th} n_e(t)\, c\, dt = \sigma_{Th} c \int_{t_p}^{0} X_e(t) n_H(t)\, dt = n_{H0} \sigma_{Th} c \int_{R_p}^{0} \frac{X_e(R)}{R^3 dR/dt}\, dR,$$

(33.4)

where σ_{Th} is the Thompson cross section for electron scattering (see Section D.1 of Appendix D and Eq. (D.1)), $X_e \equiv n_e/n_H$ is the electron fraction, and the last equality converts this to an integral over scale factor R, with $R_p \equiv R(t_p)$ its value at some past time t_p. For the simple linear expansion (empty) universe that roughly fits observations, we have $dR/dt = H_o$, with H_o the present-day Hubble constant. Using $R = 1/(1 + z)$, we can then convert this to an integral over redshift z,

$$\tau_e(z) = \tau_o \int_0^z X_e(z')(1 + z')\, dz',$$

(33.5)

where $\tau_o \equiv n_{H0} \sigma_{Th} c / H_o \approx 0.0017$ sets the overall scale of the optical depth,[1] evaluated here for $H_o \approx 70\,(km/s)/Mpc$.

To proceed, we need to determine the electron fraction X_e. This can be computed from solution of the Saha–Boltzmann ionization equilibrium discussed in Section B.2 of Appendix B. Applying Eq. (B4) to the case of pure hydrogen using $g_1/g_0 \approx 1/2$ with $n_e = n_p$, we can write

$$\frac{X_e^2}{1 - X_e} = \frac{1}{n_H(z)} \left(\frac{2\pi m_e k T(z)}{h^2} \right)^{3/2} e^{-\Delta E_H / kT(z)},$$

(33.6)

where $\Delta E_H = 13.6\,eV$ is the hydrogen ionization energy, with $T(z) = T_{cmb}(1+z)$ and $n_H(z) = n_{H0}(1 + z)^3$. Using the quadratic formula, Eq. (33.6) can be readily solved to obtain $X_e(z)$, as plotted in the top panel of Figure 33.2 for our standard cosmological

---

[1] If the universe were fully ionized today, τ_o would be the optical depth of the Hubble distance c/H_o.

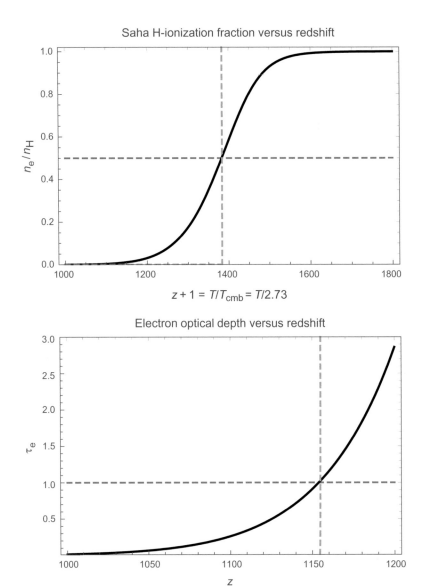

Figure 33.2 Top: Electron-to-hydrogen ratio versus redshift z, computed from solutions of the Saha–Boltzmann ionization equilibrium equation Eq. (33.6) for hydrogen. Bottom: Electron optical depth τ_e for CMB photons versus redshift z, computed from Eq. (33.5).

parameters. The dashed lines show that 50 percent ionization ($X_e = 0.5$) occurs at a redshift $z_{1/2} \approx 1380$, corresponding to a temperature $T_{1/2} \approx 3700\,\text{K}$.

The bottom panel of Figure 33.2 plots the associated redshift variation of the electron optical depth, as computed from Eq. (33.5). The dashed lines now indicate the level for unit optical depth, $\tau_e(z_{\text{rec}}) = 1$, the solution of which gives a recombination era redshift $z_{\text{rec}} \approx 1150$, corresponding now to a recombination temperature

$T_{\text{rec}} \approx 3100\,\text{K}$. The associated electron fraction $X_e = 0.012$, reflecting the fact that, for the much higher density of the recombination era, even an approximately 1 percent ionization fraction gives enough free electrons to make the radiation transport marginally optically thick.

These derived values for the redshift and temperature of the recombination agree well with the rough values assumed in the above introduction to the CMB. But they also agree remarkably well with values derived from more-complete CMB models.

33.3 Era of Nucleosynthesis

Another important constraint on the conditions in the early universe, extending to even well before the last scattering surface that formed the CMB, comes from fitting the present-day abundance of helium and other light elements. Whereas helium is synthesized in stars, it turns out that most of the helium in the universe today was actually formed in the first few minutes or so after the Big Bang, when the temperature was several billion degrees (10^9 K). This is called the "era of nucleosynthesis."

The left panel of Figure 33.3 plots the relative abundance of various light elements as function of time or temperature of the Hot Big Bang. Neutrons created at earlier, hotter times had a number ratio of about 1-to-7 to protons, and at a temperature of about 10^9 K most all these were converted into very stable helium nuclei, representing the approximately 25 percent mass fraction of primordial helium we see in the universe today. Because neutrons and protons can combine without having to

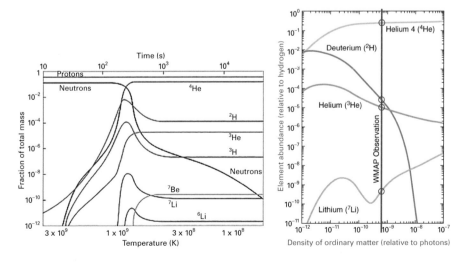

Figure 33.3 Left: Relative abundance of various light elements as function of time since the Big Bang (upper axis) or temperature (lower axis). Right: The final relative abundances as function of the number density ratio of ordinary (baryonic) matter relative to photons. Credit: NASA/WMAP.

overcome electrical repulsion, this helium production occurs quite quickly,[2] over just a few minutes!

However, it does proceed through a chain of reactions that first form the rare isotopes ^{2}H (deuterium) and ^{3}He, and so, at any given time, there are some small fractions of these. As illustrated in the right panel of Figure 33.3, the net final relative abundances of these rare isotopes depend sensitively on the number density ratio of ordinary (baryonic) matter relative to photons.

This means we can use present-day measurements of the relative abundances of helium and these rarer light elements to put a strong constraint on the matter/photon number ratio, which, as noted in Section 33.1, stays constant in time. Given the very precisely measured CMB temperature today, we can readily obtain the number density of CMB photons, thus allowing us to convert the inferred matter/photon ratio into a present-day matter density.

The upshot is that these observed relative abundances of light elements place a strong constraint on the density of baryonic matter, and its associated closure fraction Ω_b, in the present-day universe. In particular, as illustrated by the vertical line in the right panel of Figure 33.3, these measurements provide an independent check on the matter density inferred from fluctuations in the CMB measured by WMAP and Planck.

33.4   Particle Era

At even earlier epochs, with temperature $T > 10^{10}$ K, it is better to measure temperature in energy units, electron volts, instead of kelvin. Recalling that $1\,eV \approx 10^4$ K, we see that 10^{10} K ≈ 1 MeV, which is about twice the rest mass energy of an electron. At these temperatures, the photons have sufficient energy to *create* pairs of electrons and its antimatter counterpart, the antielectron, or positron. Reaction of the large number of electrons with protons then make neutrons. As the temperature cools, and the electrons and positrons annihilate, the neutron fraction freezes out at the 1-to-7 ratio (n-to-p) noted above, providing then the source conditions for later synthesis of about 25 percent of the mass into helium.

At even higher energies, $T > 10^{13}$ K≈ 1 GeV, collisions are now above the rest-mass energy of *protons*, $\sim$1 GeV, and so now create lots of protons + antiprotons. In this particle era, the universe was thus very nearly *symmetric* between matter and antimatter. But because of quantum fluctuations, for every billion anti-protons, there

---

[2] Free neutrons are unstable, with a half-life about 15 min, and so are rare in the universe today. As such, present-day production of helium in stellar cores requires overcoming the electrical repulsion between two protons, with relatively cool temperature $\sim$10^7 K that only bring the protons within a de Broglie wavelength of each other to allow quantum tunneling. But this proceeds only slowly, requiring a main-sequence lifetime of millions or even billions of years to convert the core hydrogen into helium. Cores of stars are thus relatively low-temperature "slow cookers" of helium compared to the "flash" nucleosynthesis in the first few minutes of the Hot Big Bang, made possible by the extreme temperatures $\sim$10^9 K and the earlier production of free neutrons by merging of electrons and protons.

were about a billion + one protons, from "spontaneous symmetry breaking." As the temperature cooled, each antiproton was annihilated with a pairing proton, producing the photons we see in the CMB today, with just the extra one in a billion proton left behind. The upshot is that, because of this spontaneous symmetry breaking of quantum physics, we find ourselves today in a matter universe, instead of an antimatter universe, with about a billion photons for every proton, a ratio that, as discussed in Section 33.1, remains to this day.

At even higher temperatures, protons and antiprotons are broken in to sea of "quarks." These higher temperatures are also associated with a "merging" of fundamental forces: At $T \sim 250$ GeV, electricity and magnetism merge with the weak nuclear force, giving what is called the *electroweak* force. Beyond this, it takes a *much* higher energy, $T \sim 10^{16}$ GeV (10^{25} eV $\sim 10^{29}$ K!), to merge the strong force with the electroweak force. Our best "standard model" for this is called Grand Unified Theory, or GUT, and so this merger is said to occur at the "GUT scale." By comparison, the most powerful particle accelerator we have on Earth, the Large Hadron Collider (LHC), only reaches $\sim 10^{12}$ eV (maybe extended to 10^{13} eV in the future). This means such particle colliders cannot directly test (or constrain parameters of) the GUT standard model. The unification of gravity with the GUT force occurs at an even earlier, hotter epoch, known as the Planck scale, with $T \sim 10^{20}$ GeV $\approx 10^{33}$ K.

Figure 33.4 extends the cosmic scale range shown in Figure 1.1, now adding 20 orders of magnitude below the nuclear scale, down to the Planck length. In this domain, length scales are linked to energy through the de Broglie wavelength, $\lambda \sim hc/E \approx 10^{-6}$m (eV/$E$). The GUT scale, wherein all the forces except gravity unify, occurs at energy $E_{GUT} \sim 10^{25}$ eV, corresponding to a length scale $\lambda \sim 10^{-31}$, some 16 orders of magnitude below the nuclear scale, but still 4 orders of magnitude above the Planck scale. By comparison, the Large Hadron Collider has just been able to detect the Higgs boson. With rest-mass energy of 125 GeV, this corresponds to a scale $\sim 10^{-17}$ m, just ~ 2 orders down from the nucleus, and so still 16 orders from the Planck scale. Some wonder if the domain between might not have much of interest, a kind of "size scale desert." Building colliders to find even higher-energy particles than the Higgs boson will be difficult and expensive, but extending our cosmological studies to earlier epochs with higher temperatures could provide key constraints on physics at these tiny scales.

The upshot is that the Planck era is at the very frontier, where our current physical understanding is untested and breaks down into a "quantum foam." There are competing ideas – e.g., string theory, or super-gravity – for describing this unification of gravity with the GUT force; these are grounded in elegant but purely mathematical and theoretical arguments, whose predications involve energy scales well beyond the capability of any foreseeable particle collider experiments.

Instead, the best current options for testing them may lie in cosmological studies that look for signatures of the operation of these forces in the very tiny instants after the Big Bang.

So, quite ironically, our ability to understand physics at these tiniest of scales seems intimately linked to the largest-scale studies of the universe and its origins.

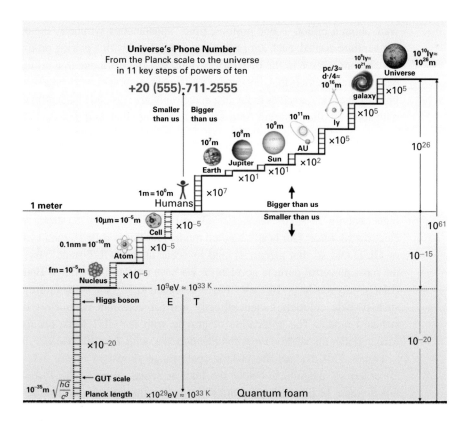

Figure 33.4 Extension of the "Universe's Phone Number" graphic in Figure 1.1, now extending 20 orders of magnitude *below* the nuclear scale, down to the Planck length, at which gravity melds with quantum physics. In this domain, length scales are linked to energy E through the de Broglie wavelength $\lambda \sim hc/E \approx 10^{-6}$ m (eV/E). The downward arrow marks the range in energy from the nuclear scale $E \approx 10^9$ eV $= 1$ GeV to the Planck scale $E \approx 10^{29}$ eV, along with associated ranges in temperature T. This extra 20 powers of ten to extend down to the Planck scale can be incorporated by now adding a "country code" 20 (which happens to be for Egypt) to the previous 10 digits, with the full number (+20-555-711-2555) now spanning the 61 powers of ten from the Planck length to the size of the observable universe.

33.5   Era of Cosmic Inflation

Despite its successes in explaining the CMB and the relative cosmic abundances of helium and other light elements, it became clear that this standard Hot-Big-Bang model could not readily explain certain quite general properties of the observed universe. These can be broken down into three fundamental problems.

1. *Flatness problem* Why is/was the total Ω so close to one, implying the universe is very "flat"? In other words, why is the total energy of the universe nearly zero (kinetic + vacuum = − gravitation)?

2. *Horizon problem* Why is the universe so isotropic, given that opposite sides of the sky should have been outside each other's "light-travel horizon" in the early universe, and thus unable to communicate to establish a common temperature?
3. *Structure problem* What is the origin of all the structure in the present-day universe, given that it started out so homogeneous? What caused the small fluctuations we have now detected in the CMB?

To answer these questions, in 1980 a young MIT physicist named Alan Guth proposed that the very early universe, at a time $<10^{-32}$ s, experienced a period of extreme, exponential *inflation*, expanding by a factor $>10^{30}$ (!) over that tiny timescale!

He speculated that this may have been powered by the energy generated in the freezing out of the GUT force from the electroweak and strong force, i.e., that it occurred toward the end of the GUT era mentioned above.

This notion of *cosmic inflation* provides potential answers to all three problems.

1. *Flatness problem* The inflation of the universe's size by a factor $\sim 10^{30}$ means that any curvature in the preinflated universe is greatly reduced, much as the curvature of a sphere is reduced by increasing its radius.
2. *Horizon problem* Because the preinflated universe was so small, sections that would end up at opposite sides of our present-day sky were initially very close together, within each other's light horizon, and thus could be causally homogenized to the nearly same properties.
3. *Structure problem* This smallness of the preinflated universe also means that, like atoms, nuclei, and elementary particles today, it was subject to "quantum fluctuations" associated with the uncertainty principal. The initially tiny physical scale of these fluctuations was amplified by inflation to much larger structures that we see today in the angular spectrum of fluctuations of the CMB. Those, in turn, were the seeds that, with the help from the extra gravitational attraction of cold dark matter (CDM), form the large-scale structure of the universe we see today.

Such explanations for these three key problems of the Hot-Big-Bang model have led to a broad (though not universal) consensus that some form of cosmic inflation did occur in the very early universe, though the details of exactly when and how it was initiated remain uncertain.

However, there are experiments underway to detect observational signatures of this inflation era. As illustrated in the top panel of Figure 33.1, the quantum fluctuations in the inflation era, which are thought to cause the fluctuations in the CMB, are also predicted to excite gravitational waves, which are like ripples in the very fabric of spacetime. Such gravitational waves can induce a *circular polarization* in the CMB radiation. The indirect detection of cosmological gravitational waves through circular polarization in the CMB would thus represent an important test of general relativity, as well as provide confirmation for, and observational constraints on, the theory of cosmic inflation.

In 2014 there were preliminary claims of such a detection from a project called Bicep2, but so far these have not been confirmed or generally accepted. Indeed, it is now generally believed that the inferred circular polarization signature was likely to be the result of contamination by foreground dust, and not the sought-after signature of gravitational waves generated by cosmic inflation. But there are hopes that with further improvements in instrumentation and data analysis, it might still be possible in the future to detect this key signature of cosmic inflation.

33.6   Questions and Exercises

Quick Questions

1. Set $X_e = 1/2$ in Eq. (33.6), and then solve iteratively to find the recombination temperature $T_{1/2}$. How does this compare to the approximate value quoted in the text?

2. Set $X_e = 0.012$ in Eq. (33.6), and then solve iteratively to find the recombination temperature T_{rec}. How does this compare to the approximate value quoted in the text?

3. Suppose at the end of the particle era, the ratio of the number of protons to neutrons had been 2-to-7, i.e. *double* the 1-to-7 it actually was.
   a. What would be the number ratio n_{He}/n_H of helium-to-hydrogen after the age of nucleosynthesis?
   b. What would be the associated helium mass fraction Y?

Exercises

1. *Age of recombination*
   a. In the simple constant expansion model of the universe $R = H_o t$ with $H_o = 70\,\text{(km/s)/Mpc}$, what is the universe's age t_{rec} (in yr) when the temperature has its recombination value $T_{rec} \approx 3300\,\text{K}$?
   b. Similarly, what is this recombination age for the flat-universe, dark-energy model with $\Omega_m = 0.3$?
   c. Which answer is closer to what you get if you google "Universe age at recombination"?
   d. Discuss the reasons for the difference. (Hint: How does the dark-energy solution (31.8) for $R(t_{BB})$ scale with time for very small times?)

2. *Near uniformity of CMB*
   For the age of the recombination $t_{rec} = 380\,000\,\text{yr}$:
   a. What is the horizon length h (in ly) light can travel in this time?
   b. For $H_o = 70\,\text{(km/s)/Mpc}$, what is the distance d (in ly) CMB light has traveled from this recombination to us today?
   c. What is the associated angle size $\alpha = h/d$ (in radians and degrees)?

 d. How is this related to the maximum coherence scale of the CMB in the absence of inflation?

 e. Discuss how an early epoch of inflation can allow this CMB coherence to extend over the whole sky.

3. *The Planck length*

According to Planck, a quantum of energy E has a de Broglie wavelength $\lambda = hc/E$. According to Einstein, such a quantum of energy has an associated mass $m = E/c^2$. Finally, according to Schwarzschild, a mass m would become a black hole if confined within a radius $r = 2Gm/c^2$.

 a. Setting $r = \lambda = \ell_{planck}$, combine these relations to solve for the associated "Planck length," ℓ_{planck}, in terms of h, G, and c. This represents the length scale at which gravity and quantum physics meld together.

 b. Evaluate this length in meters, and compare it to the characteristic length scales for a nucleus, the Higgs boson, and GUT scale cited in Figure 33.4.

 c. Compute the associated Planck energy $E_{planck} = hc/\ell_{planck}$ (in eV) and Planck temperature $T_{planck} = E_{planck}/k$.

 d. Compute the associated Planck time, $t_{planck} = \ell_{planck}/c$ (in s).

 e. For $H_0 = 70$ (km/s)/Mpc, compute $H_0 t_{planck}$, and comment on its physical significance. (Hint: How does it compare with the ratio ℓ_{planck}/d_H, where d_H is the Hubble length?)

4. *Transition from radiation-dominated to matter-dominated universe*

In the early, radiation-dominated universe, the number of photons stayed constant, but the redshift from expansion made the average energy of each photon scaled as $1/R$, meaning the mass-density equivalent of the radiative energy then scaled as $\rho_r \sim 1/R^4$.

 a. Using Eq. (30.8) with the substitution $\rho_0 \to \rho_r$, show that $\dot{R} = A/R$, where A is a constant controlled by the brightness of the early radiation.

 b. Integrate this to obtain $R(t)$ for this early time.

 c. How does the time dependence compare to that predicted by the dark-energy solution (31.8) at early times?

 d. As a function of A, find values for $R_t \equiv R(t_t)$ and time t_t for the transition from being radiation-dominated to matter-dominated.

 e. Use the result in Eq. (33.1) to estimate the values of A, R_t, and t_t.

 f. In terms of the dark-energy solution in (31.8) for R, show that the function $R_b(t) = R(t)[1 + (t_t/t)^{1/6}]$ bridges the scalings for both the radiation- and matter-dominated eras. (Figure 32.1 plots the associated temperature variation $T_{cmb}/R_b(t_{BB})$.)

5. *Final exercise: the multiverse and the nature of scientific theory*

Despite the remarkable, quantitative development of modern cosmology, there remain fundamental questions, e.g., on the nature and even the reality of inflation, and on what ultimately triggered the Big Bang. Some very famous cosmologists, including the key developer of cosmic inflation Alan Guth, have conjectured that ours could be only one of huge multitude ($>10^{100}$!) of universes, each with perhaps its own special laws and properties. This notion of the "multiverse" is grounded in theoretical studies of quantum physics, string theory, and gravity; but it has so far been largely unable to point to observations or experiments that could, even in principle, test the idea. As a final exercise for this course, do some background online research on this and related questions at the frontiers of modern cosmology. Review these, including in particular a discussion of what constitutes a physical theory, and whether this necessarily demands testable predictions. Your instructor can give you further guidelines on the expected length and breadth of this exercise.

Appendix A Atomic Energy Levels and Transitions

As a basis for the examination in Part II of how various inferred basic properties of stars from Part I can be understood in terms of the physics of stellar structure, let us next consider some key physical underpinnings for interpreting observed stellar spectra. Specifically, this appendix discusses the simple Bohr model of the hydrogen atom, while the next appendix reviews the Boltzmann description for excitation and ionization of atoms. Appendix C reviews the atomic origins of opacity, and Appendix D derives the basic equation for radiative transfer in solar and stellar atmospheres.

A.1   The Bohr Atom

The discretization of atomic energy that leads to spectral lines can be understood semiquantitatively through the simple Bohr model of the hydrogen atom. In analogy with planets orbiting the Sun, this assumes that electrons of charge $-e$ and mass m_e are in a stable circular orbit around the atomic nucleus (for hydrogen just a single proton) of charge $+e$ whose mass m_p is effectively infinite ($m_p/m_e = 1836 \gg 1$) compared to the electron. The electrostatic attraction between these charges[1] then balances the centrifugal force from the electron's orbital speed v along a circular orbit of radius r,

$$\frac{e^2}{r^2} = \frac{m_e v^2}{r}. \tag{A.1}$$

In classical physics, this orbit could, much like a planet going around the Sun, have any arbitrary radius. But in the microscopic world of atoms and electrons, such classical physics has to be modified – indeed replaced – by *quantum mechanics*.[2] Just as a light wave has its energy quantized into discrete bundles called photons, it turns out that the orbital energy of an electron is also quantized into discrete levels, much like the steps of a staircase. The basic reason stems from the fact that, in the ghostly

---

[1] The force on the left-hand side of Eq. (A.1) is written here for CGS units, for which r is in centimeter and the electron charge magnitude is 4.8×10^{-10} statcoulomb (also known as "esu"), where statcoulomb^2= erg cm = dyne cm^2. For MKS units, for which the charge is 1.6×10^{-19} coulomb, there is an additional proportionality factor $1/4\pi\epsilon_0$, where $\epsilon_0 = 8.85 \times 10^{-12}$ coulomb^2/J/m is the "permittivity of free space." For simplicity, we use the CGS form here.

[2] In the classic sci-fi flick *Forbidden Planet*, the chief engineer of a spaceship quips, "I'll bet any quantum mechanic in the space force would give the rest of his life to fool around with this gadget."

world of quantum mechanics, electrons are themselves not entirely discrete particles, but rather, much like light, can also have a "wavelike" character. In fact any particle with momentum $p = mv$ has an associated *de Broglie wavelength* given by

$$\lambda = \frac{h}{mv},$$

(A.2)

where again, h is Planck's constant.

This wavy fuzziness means an orbiting electron cannot be placed at any precise location, but is somewhat spread along the orbit. But then to avoid "interfering with itself," integer multiples n of this wavelength should match the orbital circumference $2\pi r$, implying

$$n\lambda = 2\pi r = \frac{nh}{mv}.$$

(A.3)

Note that Planck's constant itself has units of momentum times distance,[3] which represents an *angular momentum*. So another way to view this is that the electron's orbital angular momentum $J = mvr$ must likewise be quantized,

$$J = mvr = n\hbar,$$

(A.4)

where $\hbar \equiv h/2\pi$ is a standard notation shortcut. The integer index n is known as the *principal quantum number*.

The quantization condition in Eqs. (A.3) or (A.4) implies that the orbital radius can only take certain discrete values r_n, numbered by the level n,

$$r_n = n^2 \frac{\hbar^2}{m_e e^2} = n^2 r_1,$$

(A.5)

which for the ground state, $n = 1$, reduces to the "Bohr radius," $r_1 \approx 0.529$ Å $= 0.0529$ nm. More generally, this implies that most atoms have sizes of a few ångström (1 Å $\equiv 0.1$ nm).

It is also useful to cast this quantization in terms of the associated orbital *energy*. The total orbital energy is a combination of the *negative potential* energy $U = -e^2/r$, and the *positive kinetic* energy $T = m_e v^2/2$. Using the orbital force balance equation (A.1), we find that the total energy is

$$E_n = -\frac{e^2}{2r_n} = -\frac{m_e e^4}{2\hbar^2}\frac{1}{n^2} = \boxed{-\frac{E_1}{n^2} = E_n,}$$

(A.6)

where

$$E_1 \equiv \frac{m_e e^4}{2\hbar^2} = \frac{e^2}{2r_1} = 2.2 \times 10^{-11}\ \text{erg} = \boxed{13.6\ \text{eV} = E_1}$$

(A.7)

denotes the ionization (also known as binding) energy of hydrogen from the ground state (with $n = 1$). Figure A.1 gives a schematic rendition of the energy levels of

---

[3] Or also, energy × time, which when used with Heisenberg's Uncertainty Principle $\Delta E \Delta t \gtrsim h$, will lead us to conclude that an atomic state with finite lifetime t_{life} must have a finite width or "fuzziness" in its energy $\Delta E \sim h/t_{\text{life}}$. This leads to what is known as "natural broadening" of spectral lines.

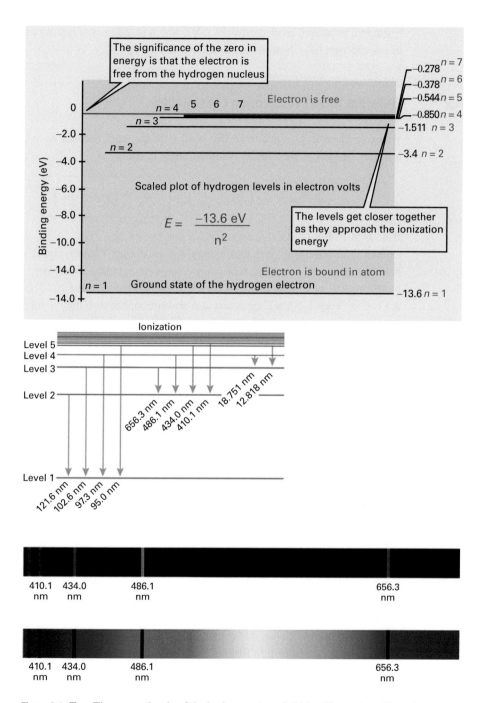

Figure A.1 Top: The energy levels of the hydrogen atom. Middle: Illustration of how the downward transitions between energy levels of a hydrogen atom give rise to emission at discrete wavelengths of a radiative spectrum. Bottom: The corresponding absorption line spectrum at the same characteristic wavelengths, resulting from absorption of a background continuum source of light that then induces *upward* transitions between the same energy levels. Figure credit: HyperPhysics project, Georgia State University, copyright Rod Nave.

hydrogen, measured in electon volts (eV), which is the energy gained when a charge of one electron falls through an electrical potential of 1 V.

A.2   Emission versus Absorption Line Spectra

When an electron changes from one level with quantum number m to another with quantum number n, then the associated change in energy is

$$\Delta E_{mn} = E_1 \left(\frac{1}{n^2} - \frac{1}{m^2} \right) = \boxed{13.6 \, \text{eV} \left(\frac{1}{n^2} - \frac{1}{m^2} \right).}$$ (A.8)

If $m > n$ in Eq. (A.8), this represents a positive energy, $\Delta E_{mn} > 0$, which can be *emitted* as a photon of just that energy $h\nu = \Delta E_{mn}$. Conversely, if $m < n$, we have $\Delta E_{mn} < 0$, implying that energy must be supplied externally, for example by *absorption* of a photon of just the right energy, $h\nu = -\Delta E_{mn}$. These processes are called "bound–bound" emission and absorption, because they involve transitions between two bound levels of electrons in an atom.

Bound–bound absorption is the basic process responsible for the absorption line spectrum seen from the surface of most stars. As illustrated in the right panel of Figure 6.2, the relatively cool atoms near the surface of the star absorb the light from the underlying layers.

On the other hand, for gas in interstellar space, the atoms are generally viewed against a dark background, instead of the bright back-lighting of a star. If the gas is dense and hot enough that collisions among the atoms occur with enough frequency and enough energy to excite the bound electrons in the atoms to some level above the ground state, then the subsequent *spontaneous decay* to some lower level will emit photons, and so result in an *emission-line spectrum*.

Recall again that Figure 6.2 illustrates the basic processes for production of emission and absorption line spectra in both the laboratory and astrophysics.

A.3   Line Wavelengths for Term Series

Instead of photon energy, light is more commonly characterized by its wavelength $\lambda = c/\nu = hc/E$. Using this conversion in Eq. (A.8), we find the wavelength of a photon emitted by transition from a level m to a lower level n is

$$\lambda_{mn} = \frac{\lambda_1}{\frac{1}{n^2} - \frac{1}{m^2}} = \frac{912 \, \text{Å}}{\frac{1}{n^2} - \frac{1}{m^2}},$$ (A.9)

where

$$\lambda_1 \equiv \frac{hc}{E_1} = \frac{h^3 c}{2\pi^2 m_e e^4} = 91.2 \, \text{nm} = 912 \, \text{Å}$$ (A.10)

is the wavelength at what is known as the *Lyman limit*, corresponding to a transition to the ground state $n = 1$ from an arbitrarily high bound level with $m \to \infty$. Of course, transitions from a lower level m to a higher level n require absorption of a photon, with the wavelength now given by the absolute value of Eq. (A.9).

The lower level of a transition defines a series of line wavelengths for transitions from all higher levels. For example, the *Lyman series* represents all transitions to/from the ground state $n = 1$. Within each series, the transitions are denoted in sequence by a lower-case Greek letter, e.g., $\lambda_{21} = (4/3)\,912 = 1216\,\text{Å}$ is called Lyman-α, while $\lambda_{31} = (9/8)912 = 1026\,\text{Å}$ is called Lyman-β, etc. The Lyman series transitions all fall in the ultraviolet (UV) part of the spectrum, which owing to UV absorption by the Earth's atmosphere is generally not possible to observe from ground-based observatories.

More accessible is the *Balmer series*, for transitions between $n = 2$ and higher levels with $m = 3$, 4, etc., which are conventionally denoted Hα, Hβ, etc. These transitions are pretty well positioned in the middle of the visible, ranging from $\lambda_{32} = 6566\,\text{Å}$ for Hα to $\lambda_{\infty 2} = 3648\,\text{Å}$ for the *Balmer limit*.

The Paschen series, with lower level $n = 3$, is generally in the infrared (IR) part of the spectrum. Still higher series are at even longer wavelengths.

Appendix B Equilibrium Excitation and Ionization Balance

B.1   Boltzmann Equation

A key issue for forming a star's absorption spectrum is the balance of processes that excite and de-excite the various energy levels of the atoms. In addition to the photon absorption and emission processes discussed above, atoms can also be excited or de-excited by collisions with other atoms. Since the rate and energy of collisions depends on the gas temperature, the shuffling among the different energy levels also depends sensitively on the temperature.

Under a condition called *thermodynamic equilibrium*, the population of electrons becomes mixed up; then if these levels were all equal in energy, the numbers in each level i would just be proportional to the number of quantum mechanical states, g_i, associated with the orbital and spin state of the electrons in that level.[1] But between a lower level i and upper level j with an energy difference ΔE_{ij}, the relative population is also weighted by an exponential term called the *Boltzmann factor*,

$$\frac{n_j}{n_i} = \frac{g_j}{g_i} e^{-\Delta E_{ij}/kT}, \tag{B.1}$$

where $k = 1.38 \times 10^{-16}$ erg/K is known as Boltzmann's constant. (Also, since energy levels are typically given in electron volt (eV), it is convenient to note that 1 eV/k $=$ 1.16×10^4 K.) At low temperature, with the thermal energy much less than the energy difference, $kT \ll \Delta E_{ij}$, there are relatively very few atoms in the more excited level j, $n_j/n_i \to 0$. Conversely, at very high temperature, with the thermal energy much greater than the energy difference, $kT \gg \Delta E_{ij}$, the ratio just becomes set by the statistical weights, $n_j/n_i \to g_j/g_i$.

As the population in excited levels increases with increased temperature, there are thus more and more atoms able to emit photons, once these excited states spontaneously decay to some lower level. This leads to an increased *emission* of the associated line transitions.

On the other hand, at lower temperature, the population balance shifts to lower levels. So when these cool atoms are illuminated by continuum light from hot layers, there is a net *absorption* of photons at the relevant line wavelengths, leading to a line-absorption spectrum.

---

[1] These orbital and spin states are denoted by quantum mechanical numbers ℓ and m, which thus supplement the principal quantum number n.

B.2   Saha Equation for Ionization Equilibrium

At high temperatures, the energy of collisions can become sufficient to overcome the full binding energy of the atom, allowing the electron to become free, and thus making the atom an *ion*, with a net positive charge. For atoms with more than a single proton, this process of *ionization* can continue through multiple stages up to the number of protons, at which point it is completely stripped of electrons. Between an ionization stage i and the next ionization stage $i + 1$, the exchange for any element X can be written as

$$X_{i+1} \leftrightarrow X_i + e^-. \tag{B.2}$$

In thermodynamic equilibrium, there develops a statistical balance between the neighboring ionization stages that is quite analogous to the Boltzmann equilibrium for bound levels given in Eq. (B.1). But now the ionized states consist of both ions, with many discrete energy levels, and free electrons. The number of *bound* states of an ion in ionization stage i is now given by something called the *partition function*, which we will again write as g_i. But to write the equilibrium balance, we now need also to find an expression for the number of states available to the *free* electron.

For this we return again to the concept of the de Broglie wavelength, writing this now for an electron with thermal energy kT. Using the relation $p^2/2m_e = \pi kT$ between momentum and thermal energy, the thermal de Broglie wavelength is

$$\Lambda = \frac{h}{p} = \frac{h}{\sqrt{2\pi m_e kT}}. \tag{B.3}$$

For each of the two electron spins, the total number of free-electron states available per unit volume is $2/\Lambda^3$. For electron number density n_e, this then implies there are $2/n_e\Lambda^3$ states for each free electron.

Using this, we can then describe the ionization balance between neighboring stages i and $i + 1$ through the *Saha–Boltzmann equation*,

$$\frac{n(X_{i+1})}{n(X_i)} = \frac{g_{i+1}}{g_i} \left(\frac{2}{n_e\Lambda^3} \right) e^{-\Delta E_i/kT}, \tag{B.4}$$

where ΔE_i is the ionization energy from stage i, and n_e is the free electron number density. The partition functions, g_i, characterizes the total number of bound states available for each ionization stage i; the large (and formally even divergent!) number of bound states can make it difficult to compute such partition functions g_i, but for hydrogen under conditions in stellar envelopes, one obtains a typical partition ratio $g_1/g_0 \approx 10^{-3}$.

Throughout a normal star, the electron state factor in parentheses is typically a huge number.[2] For example, for conditions in a stellar atmosphere, it is typically of order 10^{10}. This large number of states acts like a kind of "attractor" for the ionized state.

---

[2] As discussed in Sections 18.3, 19.2, and 19.4, it becomes order unity only in very compressed conditions, such as in the interior of a white-dwarf star, which is thus said to be *electron degenerate*.

It means the numbers in the more- versus less- ionized states can be comparble even when the exponential Boltzman factor is very small, with a thermal energy that is well below the ionization energy, i.e., $kT \approx \Delta E_i / 10$.

For example, hydrogen in a stellar atmosphere typically starts to become ionized at a temperature of about $T \approx 10^4$ K, even though the thermal energy is only $kT \approx 0.86$ eV, and thus much less than the hydrogen ionization energy $E_i = 13.6$ eV, implying a Boltzman factor $e^{-13.6/0.86} = 1.4 \times 10^{-7}$. For a partition ratio $g_1/g_0 \approx 10^{-3}$, we thus obtain roughly equal fractions of hydrogen in neutral and ionized states at modest temperature of just $T \approx 10^4$ K.

Appendix C Atomic Origins of Opacity

For solid objects in our everyday world, the interaction with light depends on the object's physical projected area, which is the source of thes concept of a "cross section." But, as noted in Section 12.3, for interstellar dust with sizes comparable to the wavelength of light, the effective cross section can depend on this wavelength, and so differ from the projected geometric area.

For atoms, ions, and electrons that make up a gaseous object like a star, the effective cross sections for interaction with light can be even more sensitive to the details. But, generally, because light is an electromagnetic (EM) wave, at the atomic level its fundamental interaction with matter occurs through the variable acceleration of charged particles by the varying electric field in the wave. As the lightest common charged particle, electrons are most easily accelerated, and thus are generally key in setting the interaction cross section. The simplest example is that of an isolated free electron, so let us begin by examining its interaction cross section and opacity.

C.1   Thomson Cross Section and Opacity for Free-Electron Scattering

As illustrated in the top left panel of Figure C.1, when a passing EM wave causes a free electron to oscillate, it generates a wiggle in the electron's own electric field, which then propagates away – at the speed of light – as a new EM wave in a new direction. Because an isolated electron has no way to store both the energy and momentum of the incoming light, it cannot by itself absorb the photon, and so instead simply scatters, or redirects it. The overall process is called *Thomson scattering*.

For such free electrons, the associated Thomson cross section can actually be accurately computed using the classical theory of electromagnetism. Intuitively, the scaling can be roughly understood in terms of the so-called "classical electron radius" $r_e = e^2/m_e c^2$, which is just the radius at which the electron's electrostatic self-energy e^2/r_e equals the electron's rest-mass energy $m_e c^2$. In these terms, the Thomson cross section for free-electron scattering is just a factor[1] 8/3 times greater than the projected area of a sphere with the classical electron radius,

---

[1] This factor 8/3 comes from detailed classical calculations, and is not easy to understand in simple intuitive terms.

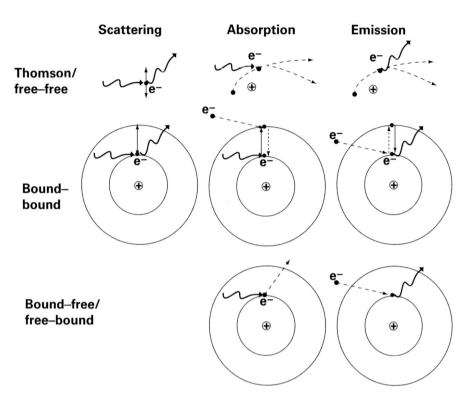

Figure C.1 Illustration of the free-electron and bound-electron processes that lead to scattering, absorption, and emission of photons. The e^- represent negatively charged electrons, while the circled $+$ represent a positively charged ion or nucleus containing one or more protons. The wavy arrowhead curves represent photons. The upward (downward) arrow lines represent excitation (de-excitation), either involving absorption or emission of radiation (solid arrows), or by collisions with electrons or ions (dashed).

$$\sigma_{\text{Th}} = \frac{8}{3}\pi r_e^2 = \frac{8}{3}\frac{\pi e^4}{m_e^2 c^4} = 0.66 \times 10^{-24}\,\text{cm}^2. \tag{C.1}$$

For stellar material to have an overall neutrality in electric charge, even free electrons must still be associated with corresponding positively charged ions, which have much greater mass. Defining then a mean mass per free electron μ_e, we can also define an electron scattering opacity $\kappa_e \equiv \sigma_{\text{Th}}/\mu_e$. Ionized hydrogen gives one proton mass m_p per electron, but for fully ionized helium (and indeed for almost all heavier ions), there are two nucleon masses (one proton and one neutron, $m_p + m_n \approx 2m_p$) for each electron. For ionized stellar material with hydrogen mass fraction X, we thus have $\mu_e = 2m_p/(1 + X)$, which then gives for the opacity,

$$\boxed{\kappa_e \equiv \frac{\sigma_{\text{Th}}}{\mu_e} = 0.2\,(1 + X)\,\text{cm}^2/\text{g} = 0.34\,\text{cm}^2/\text{g},} \tag{C.2}$$

where the last equality assumes a "standard" solar hydrogen mass fraction $X = 0.72$.

C.2   Atomic Absorption and Emission: Free–Free, Bound–Bound, Bound–Free

When electrons are bound to atoms or ions, or even just nearby ions, then the combination of the electron and atom/ion can lead to true *absorption* of a photon of light. As shown in the center top row of Figure C.1, for free electrons near ions, the shift in the electron trajectory as it passes an ion can now absorb a photon's energy, a process called *free–free absorption*. The right top panel shows that the inverse process can actually produce a photon, and so is called *free–free emission*.

The second row illustrates *bound–bound* processes, involving up/down jumps of electrons between two bound energy levels of atom, with associated absorption/emission of a photon (middle and right panel in second row), or indeed, a scattering if the absorption is quickly followed by a reemission of a photon with the same energy, but in a different direction (left panel, second row).

These bound–bound processes only work with photons with just the right energy to match the difference in energy levels, and so lead to the spectral line absorption or emission discussed earlier. But for those "just right" photons, the interaction cross section (leading to the opacity) can be much higher than for Thomson scattering or free–free absorption, because in effect it is a kind of "resonance" interaction. An everyday analogy is blowing into a whistle versus blowing just into open air. In open air, you get a weak white noise sound, made up of a range of sound frequencies/wavelengths. With a whistle, the sound is loud and has a distinct pitch, representing a resonance oscillation at some well-defined frequency/wavelength.

The third row illustrates the *bound–free* processes associated with a photon absorption that causes an atom or ion to become (further) ionized by kicking off its electron. As with electron scattering or even free–free absorption, it is a continuum (versus line) process, though it does now require that the photons have a energy equal to or greater than the ionization energy for that atom or ion. Its interaction cross section can be significantly higher than electron scattering or free–free absorption, but is generally not as strong as for bound–bound processes that lead to lines.

The cross sections, and corresponding opacities, associated with these electron + ion/atom processes are much more complicated than for free electrons, and so are difficult to cast in the kind of simple formula given in Eq. (C.2) for Thomson electron-scattering opacity. But often bound–free and free–free opacities are taken to follow so-called "Kramer's opacity,"

$$\kappa_{kr} \sim \rho\, T^{-7/2} \sim (P_{gas}/P_{rad})T^{-1/2}. \tag{C.3}$$

In stellar interiors, the ratio of gas to radiation pressure tends to be nearly constant, so that opacity decreases only weakly (as $1/\sqrt{T}$) with the increasing temperature of the interior. For the free–free case, the numerical value scales as

$$\kappa_{ff} = 0.68\, g_{ff}\, (1 - Z)(1 + X)\, \frac{P_{gas}}{P_{rad}} \sqrt{\frac{K}{T}}\, \frac{cm^2}{g} \tag{C.4}$$

$$= 1.15\, \frac{P_{gas}}{P_{rad}} \sqrt{\frac{K}{T}}\, \frac{cm^2}{g}, \tag{C.5}$$

where the latter expression takes the solar values $X = 0.72$ and $Z = 0.02$ for the mass fractions of hydrogen and metals, and assumes the simple classical approximation $g_{ff} \approx 1$ for the quantum-mechanical correction known as the "Gaunt factor." Expressions for bound-free processes are given on the Kramer's opacity Wiki page:

`https://en.wikipedia.org/wiki/Kramers'_opacity_law`

A simple rough rule of thumb is that, outside ionization zones where bound–free absorption can substantially enhance the overall opacity, stellar interiors typically have opacities that are some modest factor few times the simple electron-scattering opacity in Eq. (C.2), i.e., with a characteristic CGS value of order unity, $\kappa \approx 1 \, \text{cm}^2/\text{g}$.

Appendix D Radiative Transfer

D.1 Absorption and Thermal Emission in a Stellar Atmosphere

As noted in Section 15.2, the atmospheric transition between interior and empty space occurs over a quite narrow layer, a few scale heights H in extent, which typically amounts to about a thousandth of the stellar radius (cf. Eq. (15.5)). At any given location on the spherical stellar surface, the transport of radiation through this atmosphere can thus be modeled by treating it as a nearly *planar* layer, as illustrated in Figure D.1.

To quantify this atmospheric transition between random-walk diffusion of the deep interior to free-streaming away from the stellar surface, we must now solve a differential equation that accounts for the competition between the reduction in intensity due to *absorption* versus the production of intensity due to the local thermal *emission* $B(\tau)$. As illustrated in Figure D.1, consider a planar atmosphere with an arbitrarily large optical depth (at the bottom) seen from an observer at optical depth zero (at the top) who looks along a direction $\hat{s}$ that has a projection[1] $\mu = \cos\theta$ to the local vertical (radial) direction $\hat{r}$. The change in intensity in each differential layer dr depends on thermal emission of radiation by the local Planck function B minus the absorption of local intensity I, multiplied by the projected change in optical depth $-d\tau/\mu = \kappa\rho ds$ along the path segment ds. This leads to an "equation of radiative transfer,"

$$\mu\frac{dI(\mu,\tau)}{d\tau} = I(\mu,\tau) - B(\tau), \tag{D.1}$$

where the radial optical depth integral is now defined from a distant observer at $r \to \infty$,

$$\tau(r) \equiv \int_r^\infty \kappa\rho \, dr' \tag{D.2}$$

which thus places the observer at $\tau(r \to \infty) = 0$.

---

[1] This standard notation using μ for direction cosine here should not be confused with the notation in the previous chapters that use μ for molecular weight.

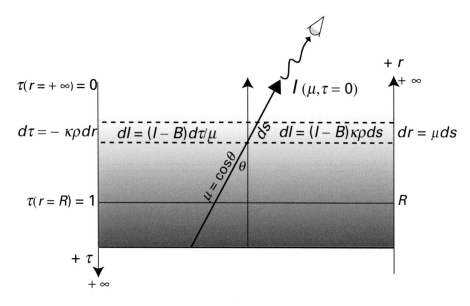

Figure D.1 Emergent intensity from a semi-infinite, planar atmosphere. Along a direction $\hat{s}$ that has projection $\mu = \hat{s} \cdot \hat{r} = \cos\theta$ to the local vertical (radial) direction $\hat{r}$, the change in intensity in each differential layer dr depends on thermal emission of radiation by the local Planck function minus the absorption of local intensity I, multiplied by the projected change in optical depth $-d\tau/\mu = \kappa\rho ds$ along the path segment ds.

D.2   The Eddington–Barbier Relation for Emergent Intensity

Equation (D.1) is a linear, first-order differential equation. By using integrating factors, it can be converted to a formal integral solution for the emergent intensity seen by an external observer viewing the atmosphere along a projection μ with the local radius

$$I(\mu, \tau = 0) = \int_0^\infty B(\tau) e^{-\tau/\mu}\, d\tau/\mu \approx B(\tau = \mu). \qquad (D.3)$$

The latter approximation here assumes the Planck function is roughly a linear function of optical depth near the star's surface, $B(\tau) \approx a + b\tau$. This so-called "Eddington–Barbier relation" states that, when you peer into an opaque radiating gas, the emergent intensity you perceive is set by the value of the blackbody function at the location of unit optical depth along that ray. This, in turn, is set by the temperature at that location, providing a more rigorous definition for what we have referred to up to now as surface brightness and surface temperature.

   An example of this Eddington–Barbier relation comes from the observed "limb darkening" of the solar disk, as illustrated by the visible-light picture of the Sun in Figure D.2 (and also in Figure 14.2). Because the line of sight looking at the center is more directly radial to the Sun's local surface, one can see into a deeper, hotter layer than from the more oblique angle when viewing toward the edge or "limb" of the solar

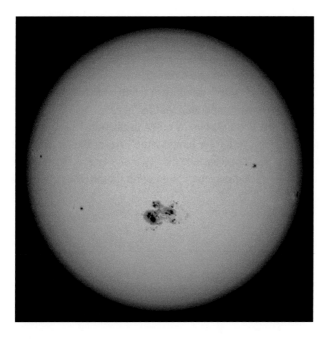

Figure D.2 Visible-light picture of the solar disk, showing the center-to-limb darkening of the surface brightness. The central lower group of dark sunspots are regions where solar magnetic storms of have inhibited the upward convective transport of heat, leading to a locally cooler, and thus darker, surface. Credit: NASA/SDO.

disk. This makes the disk appear brightest at the center, and darker as the view moves toward the solar limb.[2] The observed variation from center to limb thus provides a diagnostic of the *temperature gradient* in the Sun's surface layers.

Since stars are too far away to resolve their angular size, we cannot observe their emergent intensity $I(\mu, 0)$, but we can observe the flux $F(r) = L/4\pi r^2$ associated with the total luminosity $L = 4\pi R^2 \sigma_{sb} T_{eff}^4$. The emergent *surface* flux $F_* = L/4\pi R^2$ is obtained by integrating $\mu I(\mu, 0)$ over the 2π solid angle for the hemisphere open to empty space, giving

$$
\begin{aligned}
F_* &\equiv 2\pi \int_0^1 \mu I(0, \mu) d\mu \\
&\approx 2\pi \int_0^1 \mu B(\tau = \mu) d\mu \\
&\approx 2\pi \int_0^1 \mu(a + b\mu) d\mu \\
&= \pi B(\tau = 2/3) \\
&= \boxed{\sigma_{sb} T^4(\tau = 2/3),}
\end{aligned}
\tag{D.4}
$$

---

[2] Of course, the brightness of the Sun means we need special filters to see this effect. One should *never* look at the Sun with the naked eye.

where the third equality assumes the Planck function near the surface can be approximated as a linear function of optical depth, $B(\tau) \approx a + b\tau$.

Comparison of the final form of Eq. (D.4) with the simple discussion of surface flux in Part I of this book shows that we can identify what we have been calling the stellar "surface" as the layer where the optical depth $\tau(R) \equiv 2/3$, with the "surface temperature" likewise just the temperature at this layer.

Stars are not really blackbodies, but, as noted in Section 16.3 (see Eq. (16.8)), it is convenient to *define* a star's "effective temperature" T_{eff} as the blackbody temperature that would give the star's inferred surface flux $F_* = L/4\pi R^2$. From Eq. (D.4), we see that we can associate this effective temperature with the surface temperature at optical depth 2/3, $T_{\text{eff}} = T(\tau = 2/3)$.

Index

Printed in the United States
by Baker & Taylor Publisher Services